T0100417

# Indoor Wayfinding and Navigation

# Indoor Wayfinding and Navigation

## Hassan A. Karimi

**CRC Press**
Taylor & Francis Group
Boca Raton   London   New York

CRC Press is an imprint of the
Taylor & Francis Group, an **informa** business

CRC Press
Taylor & Francis Group
6000 Broken Sound Parkway NW, Suite 300
Boca Raton, FL 33487-2742

© 2015 by Taylor & Francis Group, LLC
CRC Press is an imprint of Taylor & Francis Group, an Informa business

No claim to original U.S. Government works

Printed on acid-free paper
Version Date: 20150128

International Standard Book Number-13: 978-1-4822-3084-0 (Hardback)

This book contains information obtained from authentic and highly regarded sources. Reasonable efforts have been made to publish reliable data and information, but the author and publisher cannot assume responsibility for the validity of all materials or the consequences of their use. The authors and publishers have attempted to trace the copyright holders of all material reproduced in this publication and apologize to copyright holders if permission to publish in this form has not been obtained. If any copyright material has not been acknowledged please write and let us know so we may rectify in any future reprint.

Except as permitted under U.S. Copyright Law, no part of this book may be reprinted, reproduced, transmitted, or utilized in any form by any electronic, mechanical, or other means, now known or hereafter invented, including photocopying, microfilming, and recording, or in any information storage or retrieval system, without written permission from the publishers.

For permission to photocopy or use material electronically from this work, please access www.copyright.com (http://www.copyright.com/) or contact the Copyright Clearance Center, Inc. (CCC), 222 Rosewood Drive, Danvers, MA 01923, 978-750-8400. CCC is a not-for-profit organization that provides licenses and registration for a variety of users. For organizations that have been granted a photocopy license by the CCC, a separate system of payment has been arranged.

**Trademark Notice:** Product or corporate names may be trademarks or registered trademarks, and are used only for identification and explanation without intent to infringe.

**Visit the Taylor & Francis Web site at**
**http://www.taylorandfrancis.com**

**and the CRC Press Web site at**
**http://www.crcpress.com**

# Contents

# Preface

In papers, reports, or websites related to navigation and transportation, the terms *wayfinding* and *navigation* are often used interchangeably. Although these terms overlap semantically, there is an important but subtle difference between them, namely, that wayfinding involves finding routes between pairs of locations whereas navigation involves traveling and receiving continuous real-time guidance while on a chosen route. Another way to observe the difference between these two terms is through common technologies used for wayfinding and navigation. Geographic information system (GIS) technology is commonly used for wayfinding, and an integration of GPS and GIS technologies (as well as others) is used for navigation. There are also differences between "wayfinding and navigation systems" and "wayfinding and navigation *services*." The former is a reference to specialized devices that provide wayfinding and navigation solutions without the need for communication links or connection to other systems and without third-party providers. The latter is a reference to services (more recently through smartphones) that provide wayfinding and navigation solutions only through wired-wireless communication links to remote systems supported by third-party providers. From a computing perspective, wayfinding and navigation systems are centralized platforms, and wayfinding and navigation services are decentralized (distributed) platforms, distributed over clients (on smartphones) and servers (maintained by providers).

Wayfinding and navigation systems and services for cars driving outdoors have been around for a long time and have become indispensable in people's mobility, in particular in unfamiliar environments. The popularity and the high demand for car wayfinding and navigation systems and services can mainly be attributed to GPS technology, which has become compact, affordable, and ubiquitous. A similar trend but a different technology (i.e., smartphones) has been the impetus for building pedestrian wayfinding and navigation systems and services. Since pedestrians seamlessly walk between different environments (outdoor to indoor and vice versa), and different data and technologies are required for indoor wayfinding and navigation, the interest in building wayfinding and navigation systems and services indoors has significantly increased, though they are still in their infancy.

While, conceptually, wayfinding and navigation in outdoor and indoor environments involve similar activities, from the physical space and functionality perspectives, there are differences between them. For example, the physical space (or environment) for pedestrian wayfinding and navigation outdoors generally involves sidewalk networks consisting of sidewalk segments and intersections, regardless of the city or the neighborhood in which wayfinding and navigation activities are needed. This is different from the physical space (or environment) for wayfinding and navigation indoors, where building structures are not uniform (e.g., different floor plans for different floors of a building and single-story vs. multistory buildings). Another difference with respect to the physical space is that while it is possible for wayfinding and navigation systems and services outdoors to provide solutions using only 2D map data, in buildings they must contain 3D (or 2.5) map data to allow travel between floors. An example of difference with respect to functionality is that unlike wayfinding and navigation systems and services outdoors, which are predominantly based on GPS as the sole positioning sensor, there is no single positioning sensor suitable for navigation in all buildings. In short, wayfinding and navigation differences indoors pose unique challenges that are not present in wayfinding and navigation outdoors.

The purpose of this book is to provide the breadth and depth of knowledge needed for understanding the issues and challenges in building wayfinding and navigation systems and services for indoors and the current approaches, techniques, and technologies considered for addressing them. This book is unique because the cognitive, positioning, mapping, and application perspectives of indoor wayfinding and navigation are discussed in one collection. This mix of different perspectives in this book helps readers better understand the issues and challenges for building indoor wayfinding and navigation systems and services, how these systems and services are different from those used outdoors, and how they can be used efficiently and effectively in challenging applications. Nonetheless, despite these unique features, to be consistent with the general perception about the terminology, as reflected in the literature, the chapters of this book sometimes reference *wayfinding* and *navigation* as they were defined previously; in other times they are used interchangeably.

The cognitive aspects of wayfinding and navigation are the subjects of the first two chapters, whereas several possible sensors for indoor positioning and a unique positioning sensor for indoor environments are discussed in chapters 3 and 4. Chapter 5 provides an example of a unique environment, while chapter 6 discusses map-aided indoor navigation. Chapters 7 through 10 focus on wayfinding and navigation issues related to the blind/visually impaired (B/VI) and discuss current approaches, techniques, and technologies for them. The last chapter addresses the privacy concerns in indoor wayfinding and navigation. The following paragraph provides the details in each chapter.

Chapter 1 discusses navigation of complex environments from the cognitive perspective. This is important in that wayfinding and navigation systems and services, like systems and services for other applications, can provide desired and reliable solutions if they are designed based on cognitive principles related to wayfinding and navigation. In chapter 2 an experiment related to indoor spatial knowledge is described. Considering that there is no sole positioning sensor that can provide acceptable positional accuracy for navigation indoors, like GPS, which has become ubiquitous for outdoor navigation, several possible positioning sensors suitable for indoor positioning are described in chapter 3. To provide a rather unique example of a positioning sensor, chapter 4 presents indoor positioning through magnetic technology and techniques. Similarly, to provide a rather unique example of the indoor environment, localization techniques and technologies in underground tunnels are presented in chapter 5. Considering that maps are a core component of wayfinding and navigation systems and services in any environment, chapter 6 discusses map-aided indoor navigation. While building indoor wayfinding and navigation systems and services poses challenges that are different from those for outdoors, there are different and unique challenges for adequately addressing the wayfinding and navigation requirements of people with disabilities. The next four chapters are devoted to the discussion and analysis of wayfinding and navigation requirements for B/VI individuals and possible approaches, techniques, and technologies for addressing them. Chapter 7 provides the challenges of indoor wayfinding and navigation for B/VI individuals. Chapter 8 discusses various suitable technologies that can be used to aid B/VI people for indoor wayfinding and navigation. Chapter 9 discusses the NavPal suite of tools that can be used to assist B/VI travelers in wayfinding and navigation in indoor environments. Chapter 10 discusses future directions in indoor wayfinding and navigation technology for B/VI travelers. Issues of privacy in indoor wayfinding and navigation systems and services are discussed in chapter 11.

This book is suitable for anyone interested in learning about approaches, maps, sensors, techniques, technologies, and applications, among other things, for indoor wayfinding and navigation. Researchers can learn about the latest research developments related to indoor wayfinding and navigation in indoor environments, developers can learn about the issues and challenges in building new indoor wayfinding and navigation systems and services, and students can learn about the fundamentals of indoor wayfinding and navigation.

# Editor

**Dr. Hassan A. Karimi** received his BS and MS in computer science and PhD in geomatics engineering. He is a professor and director of the Geoinformatics Laboratory in the School of Information Sciences at the University of Pittsburgh, Pittsburgh, Pennsylvania. His research is focused on navigation, location-based services, location-aware social networking, geospatial information systems, mobile computing, computational geometry, grid/distributed/parallel computing, and spatial databases and has resulted in more than 150 publications in peer-reviewed journals and conference proceedings, as well as in many workshops and presentations at national and international forums. Dr. Karimi has published the following books: *Advanced Location-Based Technologies and Services* (sole editor), published by Taylor & Francis Group in 2013; *Universal Navigation on Smartphones* (sole author), published by Springer in 2011; *CAD and GIS Integration* (lead editor), published by Taylor & Francis Group in 2010; *Handbook of Research on Geoinformatics* (sole editor), published by IGI in 2009; and *Telegeoinformatics: Location-Based Computing and Services* (lead editor), published by Taylor & Francis Group in 2004.

# List of Contributors

**Cristina Bahm**
University of Pittsburgh
Pittsburgh, Pennsylvania, USA

**Jörg Bankenbach**
RWTH Aachen University
Aachen, North Rhine-Westphalia,
Germany

**Sarah M. Belousov**
Carnegie Mellon University
Pittsburgh, Pennsylvania, USA

**Luigi Bruno**
German Aerospace Center (DLR)
Köln, Germany

**M. Beatrice Dias**
Carnegie Mellon University
Pittsburgh, Pennsylvania, USA

**M. Bernardine Dias**
Carnegie Mellon University
Pittsburgh, Pennsylvania, USA

**Hend K. Gedawy**
Carnegie Mellon University
Pittsburgh, Pennsylvania, USA

**Stephen C. Hirtle**
University of Pittsburgh
Pittsburgh, Pennsylvania, USA

**James Joshi**
University of Pittsburgh
Pittsburgh, Pennsylvania, USA

**Susanna Kaiser**
German Aerospace Center (DLR)
Köln, Germany

**Mohammed Khider**
German Aerospace Center (DLR)
Köln, Germany

**Prashant Krishnamurthy**
University of Pittsburgh
Pittsburgh, Pennsylvania, USA

**Adriano Moreira**
Universidade do Minho
Braga, Portugal

**Abdelmoumen Norrdine**
RWTH Aachen University
Aachen, North Rhine-Westphalia,
Germany

**Balaji Palanisamy**
University of Pittsburgh
Pittsburgh, Pennsylvania, USA

**Fernando Pereira**
European Organization for Nuclear
    Research
Geneva, Switzerland and
Universidade do Porto, Portugal

**Maria Garcia Puyol**
German Aerospace Center (DLR)
Köln, Germany

**Manuel Ricardo**
Universidade do Minho
Braga, Portugal

**Patrick Robertson**
German Aerospace Center (DLR)
Köln, Germany

**Satish Ravishankar**
Carnegie Mellon University
Pittsburgh, Pennsylvania, USA

**Samvith Srinivas**
University of Pittsburgh
Pittsburgh, Pennsylvania, USA

**Aaron Steinfeld**
Carnegie Mellon University
Pittsburgh, Pennsylvania, USA

**Ermine A. Teves**
Carnegie Mellon University
Pittsburgh, Pennsylvania, USA

**Christian Theis**
European Organization for Nuclear
    Research
Geneva, Switzerland

**George J. Zimmerman**
University of Pittsburgh
Pittsburgh, Pennsylvania, USA

# 1

# Cognition for the Navigation of Complex Indoor Environments

Stephen C. Hirtle

Cristina Robles Bahm

## CONTENTS

*Abstract:* Indoor navigation has proved to be complex to understand and to support with the use of external aids, be it signage, maps, or navigational devices. Research on the spatial cognition of complex indoor environments for the purpose of navigation is reviewed. The result of this analysis is an aggregated view of what makes indoor environments different from outdoor or transitional environments. The goal of this chapter is to coalesce the known cognitive principles to guide further research and tools in indoor wayfinding.

## 1.1 Introduction

As a catchy subtitle to his recent book *You Are Here*, Colin Ellard (2009) asked, "Why we can find our way to the moon, but get lost in the mall." He went on to describe what is known about spatial cognition from a variety of perspectives, but he discussed relatively little on indoor navigation, despite the title of the book. Navigation has often been studied in outdoor spaces, but the reality is that there is a growing literature on the nature of indoor navigation, starting almost 30 years ago. Given the large body of literature in this area, it is useful to review and categorize the knowledge gained, especially in light of recent advances in the technology for augmenting indoor locational information. The purpose of this chapter is to examine the cognitive aspect of indoor navigation, specifically in complex environments, by taking into account and synthesizing the research that has been carried out in the past three decades. While only a subset of the more than 2,000 articles published on this topic will be reviewed, it is hoped that the review will give a good background to researchers working in this area and support the material found in the rest of this book.

The work discussed falls roughly into three periods, as shown in Table 1.1. There were a number of important, well-cited, foundational studies in the 1980s that were conducted in indoor spaces and established many of the

**TABLE 1.1**

Methods by Time and Conceptual Areas for the Study of Spatial Cognition of Indoor Spaces

| Section | Subareas | Representative Articles |
| --- | --- | --- |
| 1.2 Early approaches | Spatial knowledge acquisition | Thorndyke and Hayes-Roth (1982); Gärling, Lindberg, and Mäntylä (1983) |
| | Building complexity | Moeser (1988); O'Neill (1991) |
| | You-Are-Here maps | Levine (1982) |
| 1.3 Task-oriented studies | Emergency exits | Klippel, Freksa, and Winter (2006) |
| | Modeling the navigation process | Agarwal (2005); Hirtle, Timpf, and Tenbrink (2011) |
| | Images and photographs | Ishikawa and Yamazaki (2009); Wang and Yan (2012) |
| 1.4 Cognitive-architectural perspectives | | Hölscher, Meilinger, Vrachliotis, Brösamle, and Knauff (2006) |
| 1.5 Analytical methods | Space syntax | Hillier and Hanson (1984); Richter, Winter, and Ruetschi (2009); Turner, Doxa, O'Sullivan, and Penn (2001) |
| 1.6 General principles | | Carlson, Hölscher, Shipley, and Dalton (2010); Hölscher, Montello, and Schnitzler (2013); Li and Klippel (2012) |

fundamental parameters of spatial cognition that are still used today (see Section 1.2). This was followed in the 1990s and early 2000s by a large number of empirical studies on specific indoor tasks (see Section 1.3), as well as a smaller set of papers specifically focused on the architectural attributes of buildings (see Section 1.4) or the development of specific analytic tools (see Section 1.5). Most recently, several research teams have moved away from just collecting more data and instead have focused on the development of comprehensive theories of the principles of indoor spatial cognition (see Section 1.6). The review will be followed by some thoughts on the most productive areas for future research (see Section 1.7).

## 1.2  Historical Approaches

Experimental research on cognition of indoor navigation came into its own in the early 1980s. Thorndyke and Hayes-Roth (1982) published one of the fundamental articles in this area with a focus on comparing knowledge of large indoor space (the two-wing Rand Corporation headquarters in Santa Monica), as gained from maps or from navigation. Their research established a fundamental principle that spatial information acquired from maps was fundamentally different from spatial knowledge acquired from walking through indoor spaces. The work has been generalized to other spaces beyond indoor spaces, and this early work provided the field with many of the techniques still used today.

Another set of key papers during this early time frame was from the laboratory of Tommy Gärling in Umeå, Sweden (see Gärling, 1995, for a complete review of this work). Gärling et al. (1983) wrote one of the most classic and well-cited articles on this topic, and this was one of the first articles to systematically examine familiarity, visual access, and orientation aids in finding one's way through buildings. Asking visitors to estimate the direction to interior locations was greatly facilitated by visual access between locations. However, after visitors had four tours through the building, this difference disappeared, and participants showed no difference due to visibility.

In contrast, Moeser (1988) found that in a complex indoor structure, such as a hospital, it is very difficult to learn from navigation. In a particularly surprising finding, student nurses failed to develop any sense of survey knowledge of a five-story hospital even after two years of working at the hospital. Naive participants who were asked to memorize the floor plans of the building outperformed the experienced nurses, suggesting that experience alone is not enough to gain survey knowledge when the space is complex. It is also important to note that complexity of the space is a critical parameter in testing indoor navigation.

Research during this period tended to focus either on the structure of buildings or on specific tools, such as maps and signage. In a classic article on You-Are-Here (YAH) maps, Levine (1982) argued that information, readability, and placement of YAH maps are critical to their proper use and understanding. Looking at actual YAH signs in buildings, he found that the signs are either misaligned with the environment or placed in ways that generate ambiguity. For example, rather than placing a YAH map in the center of a wall, placing it at the end of a wall in a way that is aligned with the environment will greatly facilitate the readability of the map. It is remarkable how many malls and hotels display the same generic map regardless of how it is placed in the environment (Klippel et al., 2006).

## 1.3 Task-Oriented Studies

Another interesting area of research is on tasks at hand, be it shopping or leaving in an emergency (Hajibabai, Delavar, Malek, & Frank, 2006; Raubal & Egenhofer, 1998). The problem of emergency exits is particularly acute, and it is a problem that is rarely encountered but is met with stress and limited visibility when it does happen. Klippel et al. (2006) examined one step in emergency evacuation by looking at the information that is accessible from YAH maps. The work was based on Agarwal's (2005) conceptualization of a trilateral relationship between the physical environment (the world), the schematic view (the map), and the internal representation of the space (the wayfinder). This view leads to a series of criteria for evaluating any map: completeness, perceptibility, semantic clarity, pragmatics, placement (in both global and local terms), correspondence, alignment of text, and redundancy (which has both positive and negative aspects). Klippel et al. (2006) acknowledged that information does not always have to be veridical with environment. Hirtle (2000) used the term *map gesture* to refer to general cues that can be included in the map to direct attention, in the same way a physical gesture can be helpful in telling a traveler to "head over there" without detailed route information.

In related work, O'Neill (1991) demonstrated that textual signage was as useful as graphical signage for assisting students with navigating through a campus building. However, complex buildings are difficult to navigate regardless of signage. Hirtle et al. (2011) looked at the purpose of the navigation and found that language and descriptors change in terms of both relevance and granularity depending on the task. While examining primarily outdoor navigation, Hirtle et al. (2011) argued that speeded tasks (e.g., heading to a hospital) are often described with short, direct phrases, while leisurely tasks (e.g., heading to a country inn) are often described in more verbose terms with the inclusion of various asides or observations.

Another way of examining tasks is to consider specific environments given that each environment is designed for different activities. For example, a subway station is designed to move travelers quickly and efficiently into and out of the subway system, while a large store is designed to engage the shopper and perhaps even slow the shopper down so that more goods are purchased. Looking at how individuals respond to cues upon leaving a subway station, Ishikawa and Yamazaki (2009) found that participants responded faster and more accurately to photographs that highlight subway exits rather than the same information presented on a map. Wang and Yan (2012) were able to implement an image-based app for indoor navigation based on this principle with an accuracy of 95% recognition by participants using the system.

Color has long been known to be an important cue. Evans, Fellow, Zorn, and Doty (1980) showed great improvement for recall of locations through both navigation and pointing when the interior walls of a four-story building were painted with different wall colors.

## 1.4 Cognitive-Architectural Perspectives

In the fields of architectural design and civil engineering, there has been related research that is focused on the modeling and graphic representations of the use of buildings being engineered. In addition to visualization and graphical methods (e.g., Aliaga, Rosen, & Bekins, 2007; Beneš, Šťava, Měch, & Miller, 2011), one of the key concepts is that of a "good building layout" (Bao, Yan, Mitra, & Wonka, 2013). Bao et al. (2013) argued that good building layouts can be formed as families by building a model of each building, then grouping similar buildings with similar floor plans. Rather than discuss the navigability of individual buildings, this approach suggests that navigability can be grouped into classes of buildings that are easier to navigate than other classes. In addition, the extent to which a building is prototypical for its class may make it easier to navigate. That is, a grocery store with a standard layout should be easier to navigate than, say, a two-story grocery store that would be unique in its layout.

The link between design and navigability has a rich history in the works of Lynch (1960), Stea (1974), and Passini (1992). In a recent paper using cognitive-architectural analysis, Hölscher et al. (2006) identified seven possible "hotspots" for navigational difficulties in a particular multistory conference center in Günne, Germany: The Entrance Hall, Survey Places, Floors, Dead Ends, Interior Building Structures, Public and Private Spaces, and Stairways. Each of these hotspots proved to lead to navigational difficulties. The empirical findings were supported by a space syntax analysis of the navigational bottlenecks, particularly with regard to the design of staircases in the conference center. Thus, analytic tools, such as space syntax (see next section), combined with behavioral measures, can delineate possible areas where a user could get lost when navigating a complex indoor environment.

## 1.5  Space Syntax

Space syntax (Bafna, 2003; Hillier & Hanson, 1984) has proved to be a central technique for the study of navigation and is particularly useful for indoor navigation. Space syntax is built on modeling the visibility graph of an environment (Davies, Mora, & Peebles, 2007; Turner et al., 2001). That is, by examining the intervisibility of locations using an isovist polygon, one can determine which interior locations are visible from other interior locations. Hölscher and Brösamle (2007) asked if space syntax can account for wayfinding behavior in multilevel buildings. In their experiment, 12 participants were asked to find different rooms in a large three-story campus building, which were identified by number (e.g., Room 308) or by purpose (the bowling alley in the basement). Performance was measured using six variables: (1) time to complete the task, (2) number of stops, (3) veering off a reasonable route (getting lost), (4) distance covered, (5) proportion of superfluous walking, and (6) average speed. Space syntax measures were able to account for individual performances in interesting ways. For example,

> The novices more often travel through the highly connected and integrated areas in the entry-level floor, the entrance hall and the highly integrated staircase. In terms of navigation strategies, this reflects the *central-point strategy*, most prominent among novices and least popular among experts.

Space syntax measures have also been proved useful in analyzing shopping and other spatial behaviors (Kalff, Kühner, Senk, Conroy Dalton, & Hölscher, 2010; Wiener & Franz, 2005) and modeling indoors spaces in general (Montello, 2007; Richter et al., 2009; Worboys, 2011).

## 1.6  General Principles

It has been asked if problems can be predicted in complex indoor environments by focusing on specific aspects of the navigation process and the environment itself (Li & Klippel, 2012). If a building has poor visual access, the individual's spatial ability is low, and the structure of the building is not conducive to easy navigation, can we say that people are apt to get lost there? And how can we use this knowledge to make navigation aids better? According to the current literature, these aspects of spatial cognition and the physical environment play a role in getting lost. Getting lost in buildings has also been examined directly to how ability relates to spatial learning, reasoning strategies, and individual differences among users (Carlson et al., 2010; Hölscher et al., 2013). Research has found that there are several aspects of the navigation experience and the physical building itself that can contribute to errors in navigation indoors.

The following is an aggregated list of the factors that contribute to errors in navigation for complex indoor environments (Carlson et al., 2010; Hölscher et al., 2013; Li & Klippel, 2012):

1. Visual access

2. Individual spatial ability

3. Navigation aids

4. The mental maps users construct

5. Spatial learning

6. Reasoning strategies

7. Physical environment and the structure of the building

## 1.7 Future Research Directions

With a large body of research and empirical studies having been conducted on indoor navigation, one might ask what are the remaining, emerging questions that still need to be addressed in the coming years. The analysis suggests that there are three large categories of questions that need further attention: technology, theory, and place. Each of these concepts is addressed in turn next.

### 1.7.1 Issues of Technology

Despite widespread use of global positioning technology for outdoor navigation (complete with real-time traffic information and street-view imagery), indoor navigational aids are in their infancy. As proposals are being developed for the positioning systems that work indoors (e.g., Kaemarungsi & Krishnamurthy, 2004), there are also fundamental questions of privacy and access that, for the most part transcend outdoor systems.

### 1.7.2 Issues of Theory

The proper theory for categorizing indoor spaces remains under debate. A complete theory would not only include the low-level details (e.g., Hölscher et al., 2006) but also include the high-level structures (e.g., Tversky, 1993). Just as the division of outdoor spatial knowledge into landmark, route, and survey knowledge (Werner, Krieg-Brückner, Mallot, Schweizer, & Freksa, 1997) provides a general framework, there has yet to be developed a simple higher level taxonomy for indoor spaces. Perhaps the division of simple and complex buildings or the classification into building types would be a more useful higher level framework to consider.

### 1.7.3 Issues of Place

The final question is one of place. What is an indoor space, and how is it different from an outdoor space? A recent COSIT workshop looked at the notionality of transitional spaces, such as outdoor passageways or city plazas (Kray et al., 2013). Furthermore, a building with a high percentage of restricted areas (such as a hospital) is fundamentally different from a building with a high ratio of public to private spaces (such as a shopping mall). Just as natural and built environments outdoors are fundamentally different in terms of navigation, the fundamental classes of indoor spaces are yet to be determined.

---

## 1.8 Summary

The navigation of complex indoor environments remains a burgeoning field of study with a growing number of empirical tools to monitor, model, and analyze navigational problems. As David Stea eloquently wrote in 1974, "the idea or image of a building is as important as the building itself." This powerful thought has been implemented in a variety of research projects, from observing confusion in wayfarers to using isovists and space syntax methods to model the information being extracted from the building.

The development of context-aware, adaptive navigation, be it for autonomous robots or human travelers in a new environment, is on the forefront of development (Afyouni, Ray, & Claramunt, 2014). As shown in Figure 1.1, there are environmental cues, infrastructure issues, and user preferences to bring to bear on the development of user-friendly indoor navigation systems. This chapter has focused on the user-centered context of the triad by looking at capabilities, preferences, and interface issues. The modeling technique of space syntax is but one possible modeling technique that can be used.

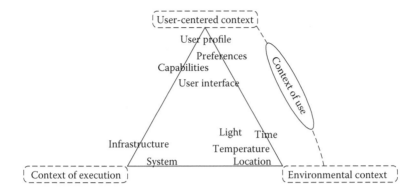

**FIGURE 1.1**
Classification of contextual dimensions for indoor navigation. Reprinted under Creative Commons Attribution 3.0 License from Afyouni, Ray, and Claramunt (2014).

Finally, one can look at the transition between indoor spaces and outdoor spaces to complete the analysis (Kray et al., 2013). Looking at passage ways, central stations, and plazas, Kray et al. (2013) found individual differences, as shown in Figure 1.2, in which some participants consider these areas to be indoor spaces in which they needed to move "into" or "out of," while others consider the same spaces to be outdoor spaces in which they needed to go "across" or "over." Kray et al. (2013) combined both behavioral and linguistic

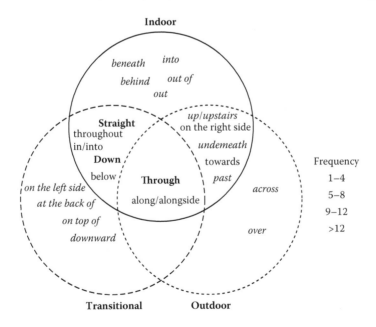

**FIGURE 1.2**
Spatial expressions relating to indoor, outdoor, and transitional spaces (from Kray et al., 2013).

analysis to develop the image of the space that travelers internalize. This kind of work suggests that difficulty in generalizing from outdoor spaces to indoor spaces and that considering the transitional spaces may provide new insights.

---

## References

Afyouni, I., Ray, C., & Claramunt, C. (2014). Spatial models for context-aware indoor navigation systems: A survey. *Journal of Spatial Information Science, 4*, 85–123.

Agarwal, P. (2005). Operationalising "sense of place" as a cognitive operator for semantics in place-based ontologies. In A. G. Cohn & D. M. Mark (Eds.), *Spatial information theory: International conference, COSIT 2005, Ellicottville, NY, USA, September 2005*. Berlin, Germany: Springer-Verlag.

Aliaga, D. G., Rosen, P. A., & Bekins, D. R. (2007). Style grammars for interactive visualization of architecture. *IEEE Transactions on Visualization and Computer Graphics, 13*(4), 786–797.

Bafna, S. (2003). Space syntax: A brief introduction to its logic and analytical techniques. *Environment and Behavior, 35*(1), 17–29.

Bao, F., Yan, D. M., Mitra, N. J., & Wonka, P. (2013). Generating and exploring good building layouts. *ACM Transactions on Graphics (TOG), 32*(4), 122.

Beneš, B., Šťava, O., Měch, R., & Miller, G. (2011). Guided procedural modeling. *Computer Graphics Forum, 30*(2), 325–334.

Carlson, L. A., Hölscher, C., Shipley, T. F., & Dalton, R. C. (2010). Getting lost in buildings. *Current Directions in Psychological Science, 19*(5), 284–289.

Davies, C., Mora, R., & Peebles, D. (2007). Isovists for orientation: Can space syntax help us predict directional confusion? In C. Hölscher, R. C. Dalton, & A. Turner (Eds.), *Space syntax and spatial cognition: Proceedings of the workshop held in Bremen, 24th September 2006* (Vol. 2, pp. 81–92). Bremen, Germany: University of Bremen.

Ellard, C. (2009). *You are here: Why we can find our way to the moon, but get lost in the mall.* New York, NY: Random House LLC.

Evans, G. W., Fellow, J., Zorn, M., & Doty, K. (1980). Cognitive mapping and architecture. *Journal of Applied Psychology, 65*, 474–478.

Gärling, T. (Ed.). (1995). *Urban cognition*. Waltham, MA: Academic Press.

Gärling, T., Lindberg, E., & Mäntylä, G. (1983). Orientation in buildings: Effects of familiarity, visual access, and orientation aids. *Journal of Applied Psychology, 68*(1), 177–186.

Hajibabai, L., Delavar, M. R., Malek, M. R., & Frank, A. U. (2006). Spatial cognition and wayfinding strategy during building fire. In *Proceedings of the 3rd international conference on spatial cognition, Rome, Italy*.

Hillier, B., & Hanson, J. (1984). *The social logic of space.* New York, NY: Cambridge University Press.

Hirtle, S. C. (2000). The use of maps, images and "gestures" for navigation. In C. Freksa, W. Brauer, C. Habel, & K. F. Wender (Eds.), *Spatial cognition II: Integrating abstract theories, empirical studies, formal methods, and practical applications* (pp. 31–40). Berlin, Germany: Springer.

Hirtle, S. C., Timpf, S., & Tenbrink, T. (2011). The effect of activity on relevance and granularity for navigation. In *Spatial information theory* (pp. 73–89). Berlin, Germany: Springer.

Hölscher, C., & Brösamle, M. (2007). Capturing indoor wayfinding strategies and differences in spatial knowledge with space syntax. In *Proceedings of 6th international space syntax symposium* (pp. 12–15).

Hölscher, C., Meilinger, T., Vrachliotis, G., Brösamle, M., & Knauff, M. (2006). Up the down staircase: Wayfinding strategies in multi-level buildings. *Journal of Environmental Psychology, 26*(4), 284–299.

Hölscher, C., Montello, D., & Schnitzler, V. (2013). *Wayfinding cognition and mobile maps for indoor settings.* GeoHCI Geographic Human Computer Interaction Workshop at CHI 2013, ACM Conference on Computer–Human Interaction, Paris, France. April 2013.

Ishikawa, T., & Yamazaki, T. (2009). Showing where to go by maps or pictures: An empirical case study at subway exits. In *Spatial information theory* (pp. 330–341). Berlin, Germany: Springer.

Kaemarungsi, K., & Krishnamurthy, P. (2004, March). Modeling of indoor positioning systems based on location fingerprinting. In *INFOCOM 2004: Twenty-third annual joint conference of the IEEE computer and communications societies* (Vol. 2, pp. 1012–1022). New York, NY: IEEE.

Kalff, C., Kühner, D., Senk, M., Conroy Dalton, R., & Hoelscher, C. (2010, August). *Turning the shelves: Empirical findings and space syntax analyses of two virtual supermarket variations.* Paper presented at the Spatial Cognition 2010 conference, Mt. Hood, OR.

Klippel, A., Freksa, C., & Winter, S. (2006). You-Are-Here maps in emergencies: The danger of getting lost. *Journal of Spatial Science, 51*(1), 117–131.

Kray, C., Fritze, H., Fechner, T., Schwering, A., Li, R., & Anacta, V. J. (2013). Transitional spaces: Between indoor and outdoor spaces. In *Spatial information theory* (pp. 14–32). Berlin, Germany: Springer.

Levine, M. (1982). You-Are-Here maps: Psychological considerations. *Environment and Behavior, 14*, 221–237.

Li, R., & Klippel, A. (2012). Wayfinding in libraries: Can problems be predicted? *Journal of Map and Geography Libraries, 8*(1), 21–38.

Lynch, K. (1960). *The image of the city.* Cambridge, MA: MIT Press.

Moeser, S. D. (1988). Cognitive mapping in a complex building. *Environment and Behavior, 20*(1), 21–49.

Montello, D. R. (2007, June). The contribution of space syntax to a comprehensive theory of environmental psychology. In *Proceedings of the 6th international space syntax symposium, Istanbul, 2007.* ITU Faculty of Architecture, Istanbul, Turkey. (pp. iv-1–iv-12).

O'Neill, M. J. (1991). Effects of signage and floor plan configuration on wayfinding accuracy. *Environment and Behavior, 23*(5), 553–574.

Passini, R. (1992). *Wayfinding in architecture* (2nd ed.). New York, NY: Van Nostrand Reinhold.

Raubal, M., & Egenhofer, M. J. (1998). Comparing the complexity of wayfinding tasks in built environments. *Environment and Planning B, 25*, 895–914.

Richter, K. F., Winter, S., & Ruetschi, U. J. (2009, May). Constructing hierarchical representations of indoor spaces. In *Mobile data management: Systems, services and middleware, 2009. MDM '09. Tenth international conference on* (pp. 686–691). New York, NY: IEEE.

Stea, D. (1974). Architecture in the head: Cognitive mapping. In J. Lang, C. Burnette, W. Moleski, & D. Vachon (Eds.), *Designing for human behavior: Architecture and the behavioural sciences*. Stroudsburg, PA: Dowden, Hutchinson and Ross.

Thorndyke, P. W., & Hayes-Roth, B. (1982). Differences in spatial knowledge acquired from maps and navigation. *Cognitive Psychology, 14*(4), 560–589.

Turner, A., Doxa, M., O'Sullivan, D., & Penn, A. (2001). From isovists to visibility graphs: A methodology for the analysis of architectural space. *Environment and Planning B, 28*(1), 103–121.

Tversky, B. (1993). Cognitive maps, cognitive collages, and spatial mental models. In *Spatial information theory: A theoretical basis for GIS* (pp. 14–24). Berlin, Germany: Springer.

Wang, E., & Yan, W. (2012). iNavigation: An image based indoor navigation system. *Multimedia Tools and Applications*, 1–19.

Werner, S., Krieg-Brückner, B., Mallot, H. A., Schweizer, K., & Freksa, C. (1997). Spatial cognition: The role of landmark, route, and survey knowledge in human and robot navigation. In *Informatik '97 Informatik als Innovationsmotor* (pp. 41–50). Berlin, Germany: Springer.

Wiener, J. M., & Franz, G. (2005). Isovists as a means to predict spatial experience and behavior. In *Spatial cognition IV: Reasoning, action, interaction* (pp. 42–57). Berlin, Germany: Springer.

Worboys, M. (2011, November). Modeling indoor space. In *Proceedings of the 3rd ACM SIGSPATIAL international workshop on indoor spatial awareness* (pp. 1–6). New York, NY: ACM.

# 2

## *The Role of Affect on Expanding Indoor Spatial Knowledge*

**Samvith Srinivas**
**Stephen C. Hirtle**

## CONTENTS

*Abstract:* Indoor spaces are fundamentally different from outdoor spaces in terms of the lack of vistas and restrictions on movement. Past literature on how individuals build indoor spatial knowledge over time through extended navigation routes is reviewed. New virtual reality studies on learning indoor spaces using either simple routes or complex routes under either motivated conditions or control conditions are presented. Participants in the motivated group were encouraged to perform the task as quickly as possible, while the control group was under no specific time constraints. Both groups were then tested on their ability to follow schematized instructions to explore unfamiliar areas in the virtual reality environment. Performance of the various spatial tasks across the motivated and control groups indicated that motivation improved performance in all but the most complex conditions. Results of the research suggest to those who design built environments and future wayfinding systems the importance of considering route complexity and knowledge and affect of the traveler.

## 2.1 Introduction

Wayfinding is a fundamental task that humans and other species are involved in on a constant basis (Golledge, 1999). The study of wayfinding behavior has a rich history in the literature, including the comparison of navigation behavior in familiar and unfamiliar environments (e.g., Streeter, Vitello, & Wonsiewicz, 1985) and the study of route directions in outdoor environments (e.g., Denis, Pazzaglia, Cornoldi, & Bertolo, 1999; Fontaine & Denis, 1999). This general line of research has examined a wayfinder's acquisition of spatial knowledge (Golledge, 1992), a wayfinder's conceptualization and internal representations of space (Mark, Freksa, Hirtle, Lloyd, & Tversky, 1999; Tversky, 1993), a wayfinder's interaction with navigation aids (Krüger et al., 2004; Streeter et al., 1985), and the importance of landmarks (Raubal & Winter, 2002; Sorrows & Hirtle, 1999; Tom & Denis, 2003), among many other areas.

Researchers have also explored the wayfinding concepts related to environments and route directions of varying complexity and granularity. Klippel and colleagues (2009) explored methods to provide *cognitively ergonomic* route directions that account for urban granularities. Tenbrink and Winter (2009) explored the nature of route directions generated by humans and systems to describe traveling through multimodal (walking and public transport) environments. Schmid (2007, 2008) explored the formulation of route descriptions that depend on an individual's spatial knowledge. In this work, Schmid automatically generated route directions that vary depending on one's level of spatial familiarity.

As a way to model wayfinding in partially familiar environments, Srinivas and Hirtle (2007) presented a theoretical approach that represents both known regions and unknown regions along a route. An example of such a route is a wayfinder traveling from her home to a new restaurant in an unfamiliar part of her town. Part of the route will be familiar from past travels, while other parts (typically the end of the route) might be unknown. Such partially familiar routes were represented as knowledge routes (*k-routes*), where each k-route consists of a series of known segments and novel segments. Furthermore, the adjacent known segments can be combined through a process called *knowledge chunking* of route direction elements. Knowledge chunking involves grouping all the segments in the region into one "knowledge chunk." These concepts for coding familiarity, and knowledge chunking, serve as a basis to generate route directions that are schematized based on a wayfinder's prior knowledge—a concept that the authors referred to as "knowledge-based schematization." The knowledge route theory of Srinivas and Hirtle (2007) formed the theoretical basis for the empirical study.

Rather than examining outdoor wayfinding as discussed by Srinivas and Hirtle (2007), the present study looks at indoor wayfinding in terms of both learned spaces and expanded spaces. Indoor spaces have received less attention in the literature than outdoor spaces but are interesting in that it is often harder for people to generalize to new unseen areas. In this study, we examine how representations are built during the initial experience with the indoor space and how the knowledge is integrated with new knowledge.

## 2.2 Learning New Spaces: Experimental Evidence

### 2.2.1 Method

To examine wayfinding in the context of knowledge route theory, researchers asked participants to navigate through an indoor virtual reality (VR) environment under one of two conditions: motivated or nonmotivated (control). In the first phase of the experiment, they learned four paths through a space, two of which were simple paths (only one turn) and two of which were complex paths (five turns). In the second phase of the experiment, they were tested on their ability to navigate along these paths. In the third phase of the study, they were tested on their ability to generalize their knowledge to new paths in the same, well-learned space. Together, these three phases made up Stage 1 of the experiment. In Stage 2 (Phase 4) of the experiment, participants were asked to navigate to new locations beyond the original space using schematized directions.

### 2.2.1.1 Participant Recruitment

Forty-two participants were recruited through flyers posted around the University of Pittsburgh campus, participants were paid $15 for their participation in the experiment that lasted between 1 and 1 and a half hours. Their ages ranged from 18 to 36 years, with a mean of 23 years. One participant was omitted from the analysis because of a misunderstanding of the instructions. Another participant was omitted from the analysis because of a lack of comfort with navigating the VR environment during testing. The resulting sample consisted of 20 female and 20 male participants.

### 2.2.1.2 Materials

The materials consisted of a standard test for working memory capacity (Smith & Kosslyn, 2007) and the Perspective Taking/Spatial Orientation Test (Hegarty & Waller, 2004). A background questionnaire and posttest questionnaire to record participant's experiences were administered. In addition, a separate questionnaire measuring the participant's confidence in locating landmarks within the learned space was given. A single projector (Epson Powerlite 730c) and a laptop (Lenovo T61) were used to present the VR environment. A standard Logitech BT96a optical wired mouse and the laptop keyboard were used for navigation control. The windows desktop screen-capturing software Hypercam v.2 was used to record the participant's movements through the VR environment.

### 2.2.1.3 VR Environments

Three VR environments were constructed, each consisting of a single floor in a building. The first VR environment included a minimal H-shaped pattern of hallways, with landmarks in the opposite corners of the space, which was used for a practice environment. The second (primary) VR environment was used in Stage 1, for both training in Phase 1 and testing in Phase 2 and Phase 3 of the experiment. It consisted of corridors and rooms with 10 unique locations, as shown in Figure 2.1. Each location, shown in Figure 2.1 using the uppercase letters, consisted of a unique shape and color on the wall. Figure 2.2 shows one such location; the white arrow in Figure 2.2 corresponds to location "B" in Figure 2.1. In the Phase 1 training environment, the invisible walls were placed in a manner that allowed the participant to take no more than one wrong turn away from the main route at any intersection along the route. The placement of invisible walls is explained in more detail in Section 2.3.2.2. In Phases 2 and 3, the invisible walls were placed in a manner that allowed the participant to take at most two wrong turns. Invisible walls in the test environment restricted exploration in areas off the main route while still allowing the traveler some degree of independence.

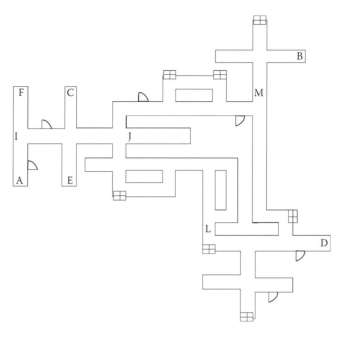

**FIGURE 2.1**
Layout of the training and test VR environments for Phase 1 and Phase 2.

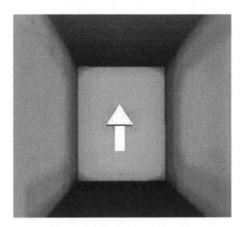

**FIGURE 2.2**
Snapshot of the white arrow that corresponds to location "B."

A third VR environment expanded the Stage 1 environment to include unknown regions, used for Stage 2 of the experiment, as shown in Figure 2.3. In this example, the origin remains the same, "A," but the new destination is point "X."

To test performance across routes of varying complexity, we ensured that the routes in the test environment satisfied certain predetermined factors.

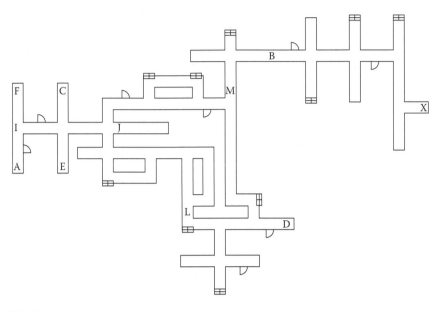

**FIGURE 2.3**
An example of an expanded test environment that was used for Stage 2.

The factors were (1) number of turns, (2) minimum number of forced views of each marked location ("I," "J," "M," and "L") during the training phase (i.e., traversal of routes 1 to 4 during the training phase ensures that each marked location "I," "J," "M," and "L" is viewed at least twice), (3) number of decision points, and (4) minimum number of alternative (longer) routes. The factors and related details are listed in Table 2.1. Routes 1 to 4 were used in the training phase and the test phase, and routes 5 to 8 were used only in the test phase. Routes 1 and 2 were the simple routes, while routes 3 and 4 were

**TABLE 2.1**

Details of Learned and Deduced Routes in Stage 1

|  | Phase 1 and Phase 2: Training and Testing | | | | Phase 3: Testing | | | |
|---|---|---|---|---|---|---|---|---|
| Route complexity | Simple | | Complex | | Simple Deduced | | Complex Deduced | |
| Route number | 1 | 2 | 3 | 4 | 5 | 6 | 7 | 8 |
| Route definition | A–J | C–I | A–B | C–D | B–M | D–L | B–E | D–F |
| Number of turns | 1 | 1 | 5 | 5 | 1 | 1 | 5 | 5 |
| Minimum number of forced views in training phase | 2 | 2 | 2 | 2 | 2 | 2 | 2 | 2 |
| Number of decision points | 2 | 1 | 7 | 8 | 1 | 2 | 6 | 9 |
| Minimum number of alternative (longer) routes | 0 | 0 | 2 | 2 | 0 | 1 | 2 | 2 |

the complex ones. Likewise, in the test session only, routes 5 and 6 were the extended simple routes, and routes 7 and 8 were the extended complex routes. The extended complex routes had the same structural complexity as a complex route; however, the task of navigating an extended complex route was estimated to be more difficult. This is because deducing the shortest path to a new location was estimated to require extra cognitive processing. Likewise, an extended simple route was estimated to require extra cognitive processing when compared to a simple route (which has similar structural complexity).

### 2.2.2 Design

Participants were assigned to either the motivated (experimental) group or the nonmotivated (control) group. In the motivated group, participants were instructed to perform the task as fast as they could. Participants were asked to imagine working under a time constraint and that time was critical. Participants were also given an estimated average time for completing the task. It was recommended to participants in this group that to be eligible for the reward, they must, at the very least, finish within the average time. The participants in this group were offered an additional reward of $15 if their performance (time to completion) ranked among the top-five best performances of all participants.

A randomized block design was used with gender as the blocking criteria. After the initial practice session, each group underwent a training phase and a test phase. The test phase consisted of two stages, first testing on the original learned routes and then testing on new deduced routes within the same space. The following subsections list details of the various phases of the experiment.

At the end of the experiment, as part of a posttest questionnaire, all participants were also asked to rate the extent to which they felt motivated, rushed, or excited during the test phase.

### 2.2.3 Procedure

#### 2.2.3.1 Practice Session

Prior to experimentation, all participants were allowed to get accustomed to the VR controls using a separate practice environment in the shape of the letter H. Participants were asked to navigate between the two landmarks placed in opposite corners without walking into the walls of the corridors. Participants were judged to be comfortable navigating within the practice environment if they made no errors (did not touch the walls of the corridors) while navigating between the two landmarks.

### 2.2.3.2 Training Phase

Upon completion of the practice session, participants in both groups underwent a training phase (Phase 1) where they were asked to navigate and learn routes within the training VR environment shown in Figure 2.1. In this phase, participants entered the test environment at a specific location (such as location "A") and were given instructions to find another location within the space identified by the object on the wall, for example, "find the White Arrow." Participants were informed that the route with the fewest turns is the shortest path. If participants did stray from the shortest path, an invisible wall blocked their progress in the wrong direction. Hence, the training environment—by design—ensured that participants followed the shortest path to an end location. The task was repeated until the participants had navigated between the two locations without deviating from the shortest path by bumping into an invisible wall. Routes 1 through 4 (see Table 2.1) were learned in this manner, with the order of routes counterbalanced across participants. Through the training process, knowledge of the four routes was established. These four routes, two simple routes and two complex routes, served as the known region. The terms *simple* and *complex* refer to the structural complexity of the route.

The training phase was followed by the spatial Perspective Taking/Spatial Orientation Test, which also served as the distractor task lasting 5 minutes.

### 2.2.3.3 Testing Phases

In the second part of the experiment, participants in both groups (motivated and control) were instructed to find the shortest path to destinations in the environment. Participants were informed that there was only one shortest path between each route. Participants were also informed that the shortest path between two locations was the route with the fewest number of turns. In the control group, participants were asked to find their destination without any time constraint and were not offered a reward for completion in quick time. In contrast, participants in the motivated group were offered a reward for quick completion and were informed that their tasks were timed (details in the following subsection).

Participants were first tested on the learned routes (Phase 2). Next, they were tested on new routes in the same environments (Phase 3). Finally, in Stage 2 of the study, they were asked to find routes in an expanded version of the space, as described later (Phase 4).

### 2.2.3.4 Measures

Measures included time to completion and number of wrong turns. A participant's movement through the VR environment was recorded using screen-capturing software. This enabled repeated playback of the route taken.

Participants indicated that they were ready to begin the wayfinding tasks by clicking on the left button of the mouse, which resulted in a flash on the screen. This was recorded as the start time. Participants were then provided with the route directions and continued to proceed with their wayfinding task. The end time was recorded the moment participants reached their target location. Thus the recorded time includes time spent reading the route directions, time spent planning the route, and time taken to move through the environment to find the target location. Hence the reaction time measured includes both the planning time and the movement time, as suggested in related work by Klatzky, Fikes, and Pellegrino (1995). A directed movement away from the shortest path into the wrong hallway was recorded as a wrong turn.

## 2.3 Results

### 2.3.1 Spatial Abilities and Learning

After participants performed two standard tests to assess any potential differences in memory or spatial skills, no differences were found. Participants' performance on the standard test for working memory capacity and the Perspective Taking/Spatial Orientation Test did not differ significantly across the motivated and control groups ($\alpha = .05$). To establish whether working memory capacity or spatial orientation ability had an effect on performance, we conducted associative analyses between working memory capacity and spatial orientation ability and wrong turn and time, for each of the four kinds of routes. No significant correlations were found. Upon completion of the training phase and prior to each test phase, participants were queried on their confidence levels in locating landmarks. A seven-point Likert item was used. There were no differences in reported confidence levels across the two groups. This implies that landmark identification across the two groups prior to each test phase was the same.

### 2.3.2 Learned Space

#### 2.3.2.1 Time

The time participants took to complete each route in Phase 2 and Phase 3 was measured. A longer task completion time indicates that participants lost their way more often, took their time in making decisions, or both. A 2 (control, motivated) × 4 (complexity: simple, simple deduced, complex, complex deduced) analysis of variance (ANOVA) revealed the main effects of experiment condition, $F(1, 36) = 4.88$, $p < .05$, indicating that the participants in the control group took a significantly longer time ($M = 30.74$) than the motivated

**TABLE 2.2**

Mean Travel Times for Simple, Simple Deduced, Complex, and Complex Deduced Routes in Stage 2

|  | Simple | Simple Deduced | Complex | Complex Deduced |
|---|---|---|---|---|
| Control |  |  |  |  |
| Mean (seconds) | 16.07 | 20.08 | 44.75 | 69.78 |
| (*SD*) | (4.66) | (5.50) | (12.56) | (21.15) |
| Motivated |  |  |  |  |
| Mean (seconds) | 12.95 | 16.13 | 35.00 | 58.88 |
| (*SD*) | (2.75) | (4.66) | (11.53) | (24.44) |
| % decrease in mean times | 28.3 | 19.7 | 21.8 | 15.6 |
| Significance | $p < .01$ | $p < .05$ | $p < .05$ | n.s. |

group ($M = 36.24$), and route complexity, $F(3, 36) = 43.32$, $p < .01$, indicating that the more complex the route, the longer the travel time. Additional $t$ tests to tease apart the source of the increased time for the control group suggested strong differences for the simple $t(37) = -3.11$, $p < .01$, simple deduced $t(37) = -2.43$, $p < .05$, and complex $t(38) = -2.58$, $p < .05$ but not for the complex deduced route, as noted in Table 2.2. Figure 2.4 displays the mean travel times of simple, simple deduced, complex, and complex deduced routes for both groups.

### 2.3.2.2 Wrong Turns

In the training phase, invisible walls were set up slightly away from each corner on the incorrect paths so that participants could start down an incorrect path but then realize the mistake. This was akin to leading someone by hand where he or she is gently nudged back after taking a wrong step. The use of invisible walls is based on the notion of virtual fixtures, introduced by Rosenberg (1993). This study used forbidden-region virtual fixtures (Okamura, 2004) where participants could see down every hallway but may be blocked from traveling by an invisible wall. In the test phase, the invisible walls were moved beyond the second wrong corner, as shown in Figure 2.5. This means that participants could make up to two wrong turns at any intersection before having to retrace their steps back to the main path. Given the complexity of the space, this ensured that they did not spend large amounts of time wandering in "back alleys," but at the same time, there would be a clear indication that participants did wander away from the appropriate path.

In Phase 2 and Phase 3, participants in all but the simplest condition (simple) exhibited wrong turns when navigating the routes. There were no significant differences in the number of wrong turns between the motivated and control groups, which was somewhat surprising. However, an analysis into the data reveals some interesting insights. Figure 2.6 shows the average

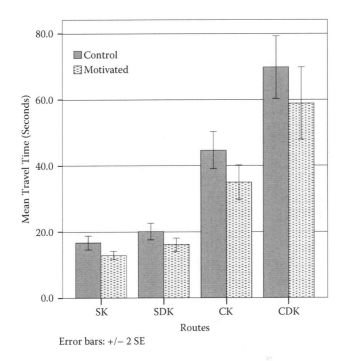

Error bars: +/− 2 SE

**FIGURE 2.4**
Plot of mean travel times (seconds) of simple (S), simple deduced (SD), complex (C), and complex deduced (CD) routes in Stage 1.

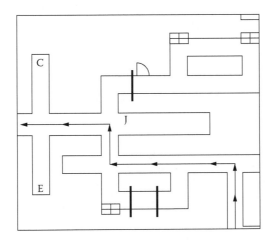

**FIGURE 2.5**
Layout of invisible walls for a route in the test phase.

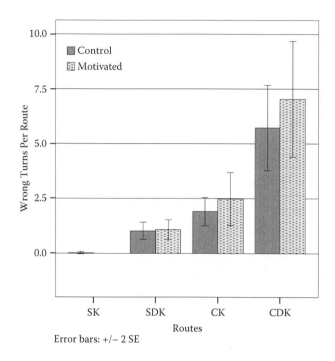

Error bars: +/− 2 SE

**FIGURE 2.6**
Plot of average number of wrong turns for each of the four types of routes: simple (S), simple deduced (SD), complex (C), and complex deduced (CD).

number of wrong turns for each of the four types of routes across the two groups. Strong positive correlations were found between wrong turns and time for motivated participants traveling the more complex routes, the complex $r(20) = .89$, $p < .01$, and complex deduced $r(20) = .85$, $p < .01$. The same correlations for the more complex routes, complex $r(20) = .45$, $p < .05$, and complex deduced $r(20) = .51$, $p < .05$, were not as strong for participants who traveled in the control condition. This suggests that participants in the motivated group spent less time making decisions, as longer travel times were the direct result of an increased number of wrong turns. In contrast, participants in the control group spent more time making decisions and less time moving, so longer travel times were often just the result of careful consideration of the next step and not necessarily indicative of a travel error. This suggests that participants in the control group were thinking more. To investigate this notion further, we conducted an analysis of the number of wrong turns that were repeated. Repeated wrong turns would suggest that participants were exploring the same space multiple times as a result of less conscious decision making. As shown in Figure 2.7, on average, participants in the motivated group had a higher number of repeated wrong turns. It is also seen that participants explored certain spaces three or four times; however, participants in the control group did not exhibit this kind of behavior.

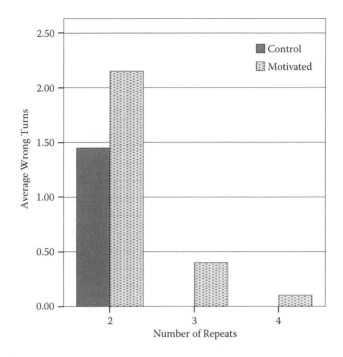

**FIGURE 2.7**
Average number of repeated wrong turns per route.

### 2.3.3 Extended Space

#### 2.3.3.1 Time

As part of the exploratory analysis in Stage 2 (Phase 4), participants were asked to follow instructions to find new targets outside the area of the earlier phases. Each path had a known (K) part and unknown (N) part. As the travel in each region could be measured independently, we first measured the time participants took to complete the known regions of the route. This time was compared to the time taken to travel the same routes in Stage 1. A 2 (control, motivated) × 4 (complexity: simple, simple deduced, complex, complex deduced) × 2 (Stage 1, Stage 2) analysis of variance (ANOVA) revealed main effects of experiment condition, $F(1, 36) = 9.35$, $p < .01$, indicating that the participants in the control group took a significantly longer time ($M = 35.10$) than those in the motivated group ($M = 29.03$); route complexity, $F(1, 36) = 171.02$, $p < .01$, indicating that the more complex the route, the longer the travel time; and stage, $F(1, 36) = 5.34$, $p < .05$, indicating that travel time was significantly faster in Stage 2 as compared to Stage 1. Additional $t$ tests to identify the source of the difference in travel times suggested that mean travel times for the simple routes were slower in Stage 2 when compared to Stage 1 for both motivated, $t(19) = -2.45$, $p < .05$, and control, $t(18) = -4.56$, $p < .01$, conditions, as noted in Table 2.3. Mean travel

**TABLE 2.3**

Mean Travel Times for Trained Regions of Stage 1 and Stage 2

|  |  |  | Simple | Simple Deduced | Complex | Complex Deduced |
|---|---|---|---|---|---|---|
| Control | Stage 1 | Mean (seconds) | 16.74 | 20.08 | 43.21 | 69.78 |
|  |  | (SD) | (4.66) | (5.49) | (10.80) | (21.15) |
|  | Stage 2 | Mean (seconds) | 23.42 | 18.55 | 48.76 | 50.17 |
|  |  | (SD) | (8.16) | (5.64) | (15.13) | (17.68) |
|  | % difference in mean times |  | 23.5 | 7.6 | 11.4 | 28.1 |
|  | Significance |  | $p < .01$ | n.s. | n.s. | $p < .01$ |
| Motivated | Stage 1 | Mean (seconds) | 12.95 | 16.12 | 35.00 | 58.88 |
|  |  | (SD) | (2.75) | (4.66) | (11.53) | (24.44) |
|  | Stage 2 | Mean (seconds) | 16.17 | 13.65 | 40.45 | 39.02 |
|  |  | (SD) | (6.90) | (6.86) | (15.01) | (12.43) |
|  | % difference in mean times |  | 19.9 | 15.3 | 13.5 | 33.7 |
|  | Significance |  | $p < .05$ | n.s. | n.s. | $p < .01$ |

times for the most complex (complex deduced) routes were faster in Stage 2 than they were in Stage 1 for both motivated $t(19) = 3.36$, $p < .01$, and control $t(19) = 4.87$, $p < .01$, conditions. The faster times perhaps reflect the difficulty of the complex deduced routes in Stage 1 of the experiment as compared to a similar task in Stage 2, where a learning effect ensured that participants were more familiar with the routes. There were no significant differences found in mean travel times of the remaining routes. Figure 2.8 and Figure 2.9 display the mean travel times in Stage 1 and Stage 2 for each route in both the control group (see Figure 2.8) and the motivated group (see Figure 2.9).

### 2.3.3.2 Wrong Turns

As mentioned earlier, in Stage 2, participants were asked to follow instructions to find new targets outside the area of Stage 1. Each path had a known (K) part and an unknown (N) part. As the travel in each region could be measured independently, we first measured the wrong turns participants made while traveling through the known regions of the route. For all the routes, the number of wrong turns per route for the known regions of Stage 2 was on par with or, in some cases, less than the wrong turns taken for the same routes in Stage 1. Figure 2.10 and Figure 2.11 display the average number of wrong turns for all routes in Stage 1 and Stage 2 for the control group (see Figure 2.10) and the motivated group (see Figure 2.11). The trend in wrong turns resembles the trend seen in the mean travel times. The most apparent difference is visible for the most complex routes. This suggests that participants made fewer errors as they learned the space across the repeated trials.

Finally, the wrong turns participants made while traveling through the unknown regions of each route were measured. This measure was used to

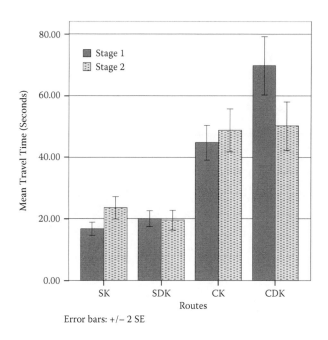

**FIGURE 2.8**
Mean times for routes in Stage 1 and Stage 2 (control group).

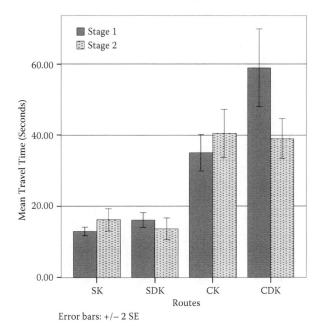

**FIGURE 2.9**
Mean times for routes in Stage 1 and Stage 2 (motivated group).

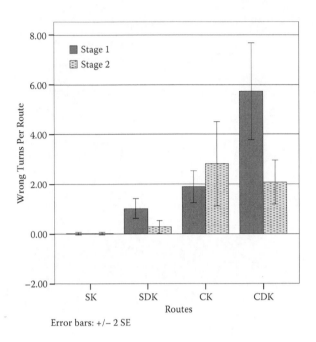

**FIGURE 2.10**
Average number of wrong turns per route for Stage 1 and Stage 2 (control group).

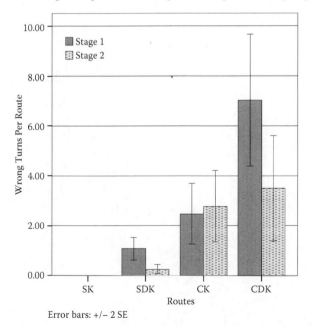

**FIGURE 2.11**
Average number of wrong turns per route for Stage 1 and Stage 2 (motivated group).

**TABLE 2.4**

Number of Wrong Turns per Decision Point

|          |                  | Simple Known | Simple Deduced Known | Complex Known | Complex Deduced Known |
|----------|------------------|--------------|----------------------|---------------|-----------------------|
| Control  | Known regions    | 0.050        | 0.167                | 0.370         | 0.467                 |
|          | Unknown regions  | 0.015        | 0.002                | 0.001         | 0.001                 |
| Motivated| Known regions    | 0.017        | 0.283                | 0.420         | 0.357                 |
|          | Unknown regions  | 0.008        | 0.008                | 0.002         | 0.003                 |

calculate an error rate. The error rate was calculated by dividing the number of wrong turns in each route by the number of decision points present for that route. The error rate was also calculated for traveling through the known regions of the route for Stage 1. Table 2.4 lists the error rates during travel across known and unknown regions. The error rate was in the range of 0.001 to 0.467 across the known and unknown regions. The error rate for unknown regions remained mostly unchanged across the routes. This, of course, is most likely due to the detailed instructions that describe traveling through these unknown sections. The error rate for traveling through known regions of the route was in the range of 0.017 to 0.467. The value for the simplest route was 0.017, and the value for the most complex route was 0.467. The varying error rate is likely due to the varying complexity of the navigation task performed. The error rates help explain the effectiveness of, and the overhead involved with, knowledge-based schematized directions.

### 2.3.3.3 Motivation

Upon completion of the experiment, we queried participants in both groups on their level of motivation and excitement and the extent to which they felt rushed. No significant differences across the two groups were found on the motivation and excitement measures. This lack of a difference could be attributed to the participants' interpretation of the query. However, participants in the motivated group reported that they felt more rushed while performing the experiment than participants in the control group. Responses to a seven-point Likert item indicate that more participants in the motivated group reported feeling rushed to extremely rushed (80%) than did participants in the control group (35%), $\chi^2 (6, N = 40) = 14.86, p = .02$.

## 2.4 Discussion

This research was designed to explore the effect of motivation on human wayfinding for indoor navigation. Specifically, Stage 1 of the study was designed to explore the effect of motivation on the time- and wrong-turn-related performance on variably complex routes in a partially known region of space. The results indicate that route complexity does in fact interact with motivation. Motivation improved time-related performance of simple and moderately complex tasks. However, motivation failed to improve time-related performance on the most complex tasks. No significant difference was found with respect to the number of wrong turns made between the two groups. Hence, motivated participants performed their tasks in less time, but they did not make fewer errors.

Earlier studies have shown that affect can improve performance of simple tasks, whereas it can hinder performance on more complex tasks (Farber & Spence, 1953; Vaughn & Diserens, 1930). The results of our study follow a similar trend. While performance in the most complex (complex deduced routes) tasks was not hindered by motivated instructions, motivated participants failed to improve their performance on these tasks. It is interesting to note that a complex deduced route and a complex route share a similar structural complexity; the only difference between these routes is the extra cognitive processing to deduce the destination in a complex deduced route. This suggests that with respect to motivated travel, structural complexity of the route may be less of a factor if the route is well known.

The results of Stage 1 of our study gain relevance as they form the basis for future work that would investigate possible performance degradation on very complex tasks, under highly motivated or rushed conditions. The results may also be viewed in the context of the real-world travel scenarios as mentioned in the introduction. The results suggest that rushed or highly motivated traveling through an unfamiliar airport will be less productive than rushed or highly motivated traveling through a familiar neighborhood park (assuming that both environments have a similar structural complexity). This was seen as a 21.8% significant difference in complex routes and a 15.6% (n.s.) difference in complex deduced routes (as shown in Table 2.2). The findings suggest that it may be important to consider route complexity when designing indoor environments. The findings are especially relevant for the design of environments that may involve traveling under stressful situations, such as hospitals, airports, fire stations, and so on. The results suggest that it may be important to design simple routes with fewer turns, especially if the routes are frequently traveled and are important or key routes within the built environment.

It is also important to compare the nature of this study, and the implications of the results, with the choking under pressure studies (Baumeister, 1984; Beilock & Carr, 2001; Lewis & Linder, 1997; Wine, 1971). The nature of affect induced in the motivated group varies, but only slightly, from the performance under pressure studies. In the present research, participants in the motivated group were motivated by reward and rushed because of an imposed time constraint. While this pressure is subtly distinct from the performance pressure induced as part of the choking studies, the one common aspect of the two kinds of manipulations is an affective state of high arousal, or pressure, that demands a higher than usual task performance. The nature of the task performed also plays a vital role in the pressure-related performance degradation. Explicit monitoring theories usually explain degradation of sensorimotor tasks such as golf putting (Beilock & Carr, 2001; Lewis & Linder, 1997) or the "roll up" game (Baumeister, 1984) that are well learned. Distraction theory better explains performance degradation in tasks that involve access to elements in working memory. The tasks presented as part of this research primarily involved access to stored knowledge but also involved some amount of sensorimotor control (control of VR). Given that the major aspect of the task performed by participants in this research involved access to stored memory, the wrong-turn-related performance degradation may be primarily explained by the distraction theory. However, as Beilock and Carr (2001) mentioned, it is likely that the two theories may complement rather than oppose each other. This is perhaps true for more complex tasks such as human wayfinding, where both theories may be needed to completely and accurately explain performance degradation.

In Stage 2, we examined what happens when the space is expanded beyond the area that participants feel they know well. This is akin to having to explore new corridors in a hospital or campus building, where the structure does not mimic the known corridors.

In general, participants in both stages were able to use schematized directions and comfortably complete their wayfinding tasks. Trends in the data indicate that the effectiveness of schematized directions was found to be inversely proportional to the complexity of the task at hand. Schematized directions were found to be most effective for the simplest wayfinding task. This is of importance considering that participants traveling through unknown regions of the route were guided by detailed route directions, and those traveling through the known regions had no detail. These directions were schematized based on the knowledge-chunking concepts mentioned earlier. This suggests that knowledge-based schematized directions may be presented if individuals have some knowledge of the environment and the task at hand is relatively simple or if knowledge is firmly established while people perform complex wayfinding tasks.

The effect of motivation on the performance of wayfinding tasks gains importance in light of the recent research in automatic route guidance systems. Recent efforts in this area involved modeling user's knowledge with the goal of tailoring output of route directions or maps to the user's mental representations or prior route or survey knowledge (Patel, Chen, Smith, & Landay, 2006; Schmid, 2008; Schmid & Richter, 2006; Srinivas & Hirtle, 2007; Tomko & Winter, 2006). These personalized routes, while describing traveling through a known region, consist of simpler directions with fewer route direction elements. This in turn reduces the cognitive load of the wayfinder. These research initiatives are an important step toward the automatic generation of personalized routes based on user familiarity.

The results of this research are relevant for future automated route guidance systems that may need to tailor their personalized route directions in accordance with not only a wayfinder's prior knowledge but also their affective state. This would require a future wayfinding system to sense a wayfinder's affective state and route knowledge and present him or her with personalized route guidance based on these factors. Recent research in this area suggests that detecting stress levels of a driver in driving conditions might prove useful in customizing the driver's "in-vehicle environment." Lin and colleagues (2007) made a small but important technical step in precisely this direction by developing a "smart wheel." The device was shown to satisfactorily measure a driver's pulse wave, breathing wave, skin temperature, and gripping force in real time. Lin and colleagues stated that such a system would prove useful in enhancing driver safety. While their study was intended to improve driver safety, it is easy to imagine that these systems could also be used to improve usability of indoor wayfinding systems. One can imagine future wayfinding systems that will direct highly motivated or stressed users to longer but simple routes with fewer turns.

## Acknowledgments

This chapter is based on the doctoral dissertation of the first author, submitted to the University of Pittsburgh, and was supported by a dissertation research award from Dean Ron Larsen at the School of Information Sciences, University of Pittsburgh. The study was guided by the direction of committee members Peter Brusilovsky, Alexander Klippel, C. Michael Lewis, and Christian Schunn.

## References

Baumeister, R. F. (1984). Choking under pressure: Self-consciousness and paradoxical effects of incentives on skillful performance. *Journal of Personality and Social Psychology, 46*(3), 610–620.

Beilock, S. L., & Carr, T. H. (2001). On the fragility of skilled performance: What governs choking under pressure? *Journal of Experimental Psychology: General, 130*(4), 701–725.

Denis, M., Pazzaglia, F., Cornoldi, C., & Bertolo, L. (1999). Spatial discourse and navigation: An analysis of route directions in the city of Venice. *Applied Cognitive Psychology, 13*(2), 145–174.

Farber, I. E., & Spence, K. W. (1953). Complex learning and conditioning as a function of anxiety. *Journal of Experimental Psychology: General, 45*(2), 120–125.

Fontaine, S., & Denis, M. (1999). The production of route instructions in underground and urban environments. In C. Freksa & D. M. Mark (Eds.), *Spatial information theory: Cognitive and computational foundations of geographic information science* (pp. 83–94). Heidelberg, Germany: Springer.

Golledge, R. G. (Ed.) (1999). *Wayfinding behavior: Cognitive mapping and other spatial processes.* JHU Press.

Hegarty, M., & Waller, D. (2004). A dissociation between mental rotation and perspective-taking spatial abilities. *Intelligence, 32,* 175–191.

Klatzky, R. L., Fikes, T. G., & Pellegrino, J. W. (1995). Planning for hand shape and arm transport when reaching for objects. *Acta Psychologica, 88*(3), 209–232.

Klippel, A., Hansen, S., Richter, K. F., & Winter, S. (2009). Urban granularities—a data structure for cognitively ergonomic route directions. *GeoInformatica, 13*(2), 223–247.

Krüger, A., Butz, A., Stahl, C., Wasinger, R., Steinberg, K., & Dirschl, A. (2004). *The connected user interface: Realizing a personal situated navigation system.* Paper presented at the IUI2004: International Conference on Intelligent User Interfaces. Madeira, Funchal, Portugal, January 2004.

Lewis, B. P., & Linder, D. E. (1997). Thinking about choking? Attentional processes and paradoxical performance. *Personality and Social Psychology Bulletin, 23,* 937–944.

Mark, D., Freksa, C., Hirtle, S., Lloyd, R., & Tversky, B. (1999). Cognitive models of geographical space. *International Journal of Geographical Information Science, 13*(8), 747–774.

Okamura, A. M. (2004). Methods for haptic feedback in teleoperated robot-assisted surgery. *Industrial Robot: An International Journal, 31*(6), 499–508.

Patel, K., Chen, M. Y., Smith, I., & Landay, J. A. (2006). *Personalizing routes.* Paper presented at the Proceedings of the 19th annual ACM Symposium on User Interface Software and Technology. Montreux Switzerland.

Raubal, M., & Winter, S. (2002). Enriching wayfinding instructions with local landmarks. In M. J. Egenhofer & D. M. Mark (Eds.), *Geographic information science: Lecture notes in computer science* (Vol. 2478, pp. 243–259). Berlin, Germany: Springer.

Rosenberg, L. B. (1993). *Virtual fixtures as tools to enhance operator performance in telepresence environments.* Paper presented at the Telemanipulator Technology and Space Telerobotics, Boston, MA.

Schmid, F. (2007). Formulating, identifying and analyzing individual spatial knowledge. *Seventh IEEE International Conference on Data Mining Workshops* (pp. 655–660). New York, NY: IEEE.

Schmid, F. (2008). Knowledge-based wayfinding maps for small display cartography. *Journal of Location Based Services, 2*(1), 57–83.

Schmid, F., & Richter, K.-F. (2006). *Extracting places from location data streams.* Paper presented at the UbiGIS 2006 Second International Workshop on Ubiquitous Geographical Information Services, Muenster, Germany.

Smith, E., & Kosslyn, S. (2007). *Cognitive psychology: Mind and brain* (1st ed.). Upper Saddle River, NJ: Pearson Prentice Hall.

Sorrows, M. E., & Hirtle, S. C. (1999). The nature of landmarks for real and electronic spaces. In C. Freksa & D. M. Mark (Eds.), *Spatial information theory: Cognitive and computational foundations of geographic information science* (pp. 37–50). Heidelberg, Germany: Springer.

Srinivas, S., & Hirtle, S. C. (2007). Knowledge based schematization of route directions. In T. Barkowsky, M. Knauff, G. Ligozat, & D. R. Montello (Eds.), *Spatial cognition V* (Vol. 4387, pp. 346–364). Bremen, Germany: Springer.

Streeter, L. A., Vitello, D., & Wonsiewicz, S. A. (1985). How to tell people where to go: Comparing navigational aids. *International Journal of Man-Machine Studies, 22*(5), 549–562.

Tenbrink, T., & Winter, S. (2009). Variable granularity in route directions. *Spatial Cognition and Computation, 9*(1), 64–93.

Tom, A., & Denis, M. (2003). Referring to landmark or street information in route directions: What difference does it make? In *Spatial information theory: Foundations of geographic information science* (pp. 362–374). Heidelberg, Germany: Springer.

Tomko, M., & Winter, S. (2006). Recursive construction of granular route directions. *Journal of Spatial Science, 51*(1), 101–115.

Tversky, B. (1993). Cognitive maps, cognitive collages, and spatial mental models. In *Spatial information theory: A theoretical basis for GIS* (pp. 14–24). Heidelberg, Germany: Springer.

Vaughn, J., & Diserens, C. M. (1930). The relative effects of various intensities of punishment on learning and efficiency. *Journal of Comparative Psychology, 10*(1), 55–66.

Wine, J. (1971). Test anxiety and direction of attention. *Psychological Bulletin, 76*(2), 92–112.

# 3

## Technologies for Positioning in Indoor Areas

Prashant Krishnamurthy

**CONTENTS**

*Abstract:* Unlike outdoor areas where the global positioning system dominates as the positioning technology of choice for navigation, indoor areas do not have a preferred positioning technology choice for navigation. In this chapter, we review some indoor positioning technologies, such as Wi-Fi, Bluetooth, RFID (radio-frequency identification), dead reckoning, and acoustic technologies for positioning in indoor areas. We describe these technologies and discuss their suitability for indoor navigation.

## 3.1 Introduction

Positioning* in outdoor areas, especially for navigation, is well understood, with the global navigation satellite system (GNSS), to which the global positioning system (GPS) belongs, providing the backbone for positioning. In recent years, GNSS, mainly GPS, has been augmented by wireless local area networks (WLANs) based on Wi-Fi for improved positioning, especially with the emergence of smartphones. Indoor areas pose different, often challenging, problems for positioning. In terms of radio signal propagation, signals from satellites do not penetrate buildings adequately to provide positioning services. Furthermore, multipath propagation inside buildings causes problems with correctly determining the time or direction of arrival of signals. In terms of accuracy, indoor areas require more granularity for many applications of interest. GPS works well in outdoor areas where vehicles may not need average accuracy values better than 10 m for navigational purposes. However, in indoor areas, for navigation, it is likely that the accuracy has to be on the order of a meter or better. On the other hand, the speed of movement of human beings in indoor areas is much slower, and humans have a variety of options for locating themselves in indoor areas, such as using signage landmarks. However, in large indoor areas such as airports and malls, humans need positioning and navigational aids for several reasons. These reasons include finding the shortest (or easiest) path from one location to another, finding resources (such as restrooms, specific stores, airport gates, library shelves for a given call number), and in general situating themselves in their surroundings simply for awareness. A final challenge with positioning in indoor areas crops up as a result of the complications arising due to the presence of multiple floors in large buildings. While people may be able to locate themselves as being on a certain floor of a building, a positioning method and localization scheme may be rendered futile if it cannot automatically recognize this information.

As mentioned previously, positioning indoors has always been problematic since commonly used positioning infrastructures for outdoors, such as GPS, do not work very well in buildings. While GPS may indicate that a mobile device is *at a building*, the limited accuracy and available map information cannot pinpoint the exact location of the device, such as whether it is inside the building or outside the building, and if it is inside the building, where exactly it is located (e.g., on which floor). The coordinate system in indoor areas is not universally standardized. In most indoor positioning applications, local coordinates or relative coordinates are employed to position a device or human being.

---

* We use the terms *positioning, localization,* and *location estimation* interchangeably although there are differences in their meaning in various works. This is for simplicity and because the granularity of locations or coordinates of positions are not standardized.

Building an entirely new infrastructure for positioning in indoor areas would incur substantial cost in a variety of ways. The cost would include embedding hardware capabilities in mobile devices to sense signals for positioning and installing anchor devices similar to base stations at known locations to transmit signals that can be used for positioning. In the case of RF-based positioning, the cost would include the cost of radio spectrum, if it would be used solely for positioning purposes. This substantial cost of new infrastructure has led to the investigation of commonly employed RF wireless technologies for positioning purposes in indoor areas. The advantage here is that such technologies already have the available spectrum and the communication infrastructure in place. While they may be repurposed or reused for positioning, they are often not optimally designed for positioning applications. In addition to RF signals, there have been studies that have considered the use of acoustic signals, ultrasound, and dead reckoning for positioning in indoor areas.

In this chapter, we review some popular indoor positioning technologies, such as Wi-Fi, Bluetooth, RFID, dead reckoning, and acoustic technologies for positioning in indoor areas. Where possible, we discuss their suitability for indoor navigation. The reader is referred to tutorial articles on this topic such as Deak, Curran, and Condell (2012) and Gu, Lo, and Niemegeers (2009) for a complementary treatment of the subject. The literature in Deak et al. is devoted to specific implementations of positioning, while this chapter provides a more general view of indoor positioning technologies.

Figure 3.1 shows a classification of the various technologies. We can break down the technologies into RF-based and non-RF-based technologies. Under RF-based technologies, it is common to employ Wi-Fi as the technology of choice, although Bluetooth, RFID, and cell phone technologies (cellular) are also possible choices. We discuss Wi-Fi in more detail, as it has received the most attention in the literature. Among non-RF-based technologies, acoustic technologies that use either ultrasound or sound for localization have

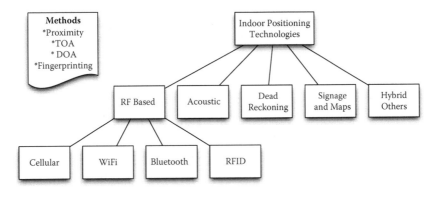

**FIGURE 3.1**
Positioning technologies and methods for indoor areas.

received attention. Dead reckoning and signage are the obvious choices for localization in indoor areas without a positioning infrastructure. There are approaches that employ more than one technology (e.g., a combination of RF and dead reckoning) as well.

Different technologies use different approaches or methodologies for locating a user or device. As shown in Figure 3.1, the common methodologies are those based on proximity to a known device, those that use the time of arrival (TOA) of a signal, those that use TOA and direction of arrival (DOA), and finally those that employ location fingerprinting. We discuss these as necessary with the technologies for positioning in subsequent sections.

Our objective in this chapter is to provide a brief summary of these technologies rather than an in-depth treatment of each approach and the various implementations, prototypes, and performance evaluations. The reader is encouraged to explore the references and the references within the references for more details.

## 3.2 RF-Based Positioning Technologies

In this section, we discuss the most popular RF-based technologies for indoor positioning. We start with Wi-Fi and then discuss Bluetooth, RFID, and cell-phone-based technologies.

### 3.2.1 Wi-Fi-Based Positioning

The widespread use of Wi-Fi (Perahia & Stacey, 2008) in homes, hotels, coffee shops, airports, malls, and other large and small buildings makes Wi-Fi an attractive technology for positioning. Typically, a Wi-Fi system consists of fixed access points (APs) that are installed in convenient locations indoors. The locations of APs are usually known to the system or network administrator. Mobile devices such as laptops, cell phones, and cameras that are Wi-Fi enabled communicate with other devices or the Internet through these APs. Thus, this makes Wi-Fi suitable for positioning in addition to enabling communication for mobile devices. Wi-Fi has been standardized by the Institute of Electrical and Electronics Engineers (IEEE), and various parts of the standard fall under the IEEE 802.11 moniker. The earliest IEEE 802.11 standard was approved in 1997 and included not only RF-based communications but also diffused infrared communications. Since then, RF-based communications in the 2.4 GHz and 5 GHz bands have prevailed commercially. The IEEE 802.11 standard has evolved over time to include multiple antennas (enabling multiple-input multiple-output or MIMO technology) and bonding of frequency channels to increase data rates and throughput. The latest standard as of this writing is called the IEEE 802.11ac standard that supports data rates of several 100's of Mbps.

Wi-Fi signals, irrespective of the evolution in the physical layer, were not designed for positioning as the main application. They were primarily designed for high-speed tetherless Internet access. The harsh radio propagation environment in the indoors makes it difficult and challenging to employ well-known algorithms based on time or time difference of arrival (TOA/TDOA) to determine distances to known APs. Using direction of signals for positioning is also difficult for the same reasons in indoor areas. Multipath propagation implies that signals may arrive from directions that are very different from the source transmitter. Moreover, installing directional antennas in every Wi-Fi network is also costly. Consequently, the two approaches that have been widely considered in recent years for Wi-Fi-based positioning are those that are based on *proximity* and those that are based on *location fingerprinting*.

*Proximity-based Wi-Fi positioning:* The general idea behind proximity-based positioning is that the device is located at the location that corresponds to the nearest location of a known device. Let us suppose that a known device is at coordinates $(x_{known}, y_{known})$ in two dimensions. If this is the *closest* known device to an unknown user or mobile device to be localized, then the assumption is that this is also the estimate of the position of the unknown device. Clearly, the maximum error in positioning corresponds to the maximum possible distance from the known device at which the positioning algorithm would determine that the known device is closer than other known devices. In the case of Wi-Fi and other RF-based positioning systems, this would depend on the characteristics of the *received signal strength* or RSS from known APs in the indoor area of interest. If the accuracy is to be improved, the range of a known device has to be small. But this would require the installation of a large number of small-range devices, which could be a costly effort. These trade-offs are not specific to Wi-Fi but shared with other proximity-based schemes that may use Bluetooth, RFID, or even acoustic signals.

Computing the RSS is a requirement for most wireless communications for operational purposes. Most Wi-Fi network interface cards (NICs) are able to measure the RSS from multiple APs, albeit one at a time. Most wireless systems need the RSS information to assess the quality of the link, to make hand-offs, to adjust the transmission rates, and for other operational reasons. If there is exactly one transmitter, the average RSS from that particular transmitter in dB falls linearly with the logarithm of the distance $d$ (usually in meters) from the transmitter (Pahlavan & Krishnamurthy, 2013). That is, in the simplest case, it is possible to express the average RSS as follows:

$$RSS = P_t - K - 10\alpha \log_{10} d. \tag{1}$$

The factor affecting the slope, $\alpha$, of this linear drop is called the path-loss exponent. The transmit power is $P_t$, and $K$ is a constant that depends on the frequency and environment. The RSS can thus be used to extract the distance of the mobile device from the AP or base station in the network. This begs the question as to whether the extracted distance can be used for trilateration or triangulation of the mobile device. Unfortunately, there are significant variations around this average due to environmental effects (this is called *shadow fading*). Consequently, the errors in positioning using the extracted distance are likely to be very high, precluding this as a good solution for indoor positioning.

For the same reason, proximity-based approaches tend to have large errors. For an illustration of the challenges, consider the following example. If the transmit power of a known AP is $P_t = 30$ dBm, $K = 40$ dB, and $\alpha = 3$, the RSS is $-10 - 30 \log_{10} d$. The minimum required RSS for communications with Wi-Fi is around $-80$ dBm. This implies that the farthest distance $d$ that a Wi-Fi device may be from the known AP is $d = 215$ m from Equation 1. The base of the John Hancock Center in Chicago has dimensions of roughly 80 m by 50 m (see Skyscraperpage.com, http://skyscraperpage.com/cities/?buildingID = 17). So, if a Wi-Fi AP is located in the ground floor of the Hancock tower, an unknown device may be located only within that floor, and a more accurate position would not be possible, simply based on proximity to the AP.

If there are multiple Wi-Fi APs that are "visible" to the unknown device, it is common to assume that the one with the larger RSS is closer to the device. However, shadow fading may create the impression that an unknown device is closer to a known AP that is physically farther away than a second known AP. For example, a thick brick wall or a metallic vending machine may attenuate the RSS of a closer AP compared to the RSS from an AP that is farther away. This will lead to larger errors in positioning. As an example, consider Figure 3.2, which shows a 40 m × 30 m rectangular indoor area with a wall in between. In this figure, there is an AP that is located at the local coordinates (10,20) where the origin is at the bottom left corner of the room. Using Equation 1, we plotted the RSS contours in this region. It is clear that the wall, which is assumed to introduce a loss of 5 dB, creates a shadowed region. A mobile that is in the shadowed region, say at (40,20), may observe that the RSS from a much farther AP, say at (30,0), is higher at $-50.5$ dBm than the RSS of $-54$ dBm from an AP located at (10,20). Note that physically, the mobile is closer to the AP at (10,20) at a distance of 20 m than the AP at (30,0), which is at a distance of 22.4 m. Although this is a simplified "toy" example, it illustrates the problems with proximity-based positioning.

Therefore, we can conclude that the large range of Wi-Fi transmissions is a potential drawback for proximity-based positioning with Wi-Fi. One possible solution to this problem is to use many inexpensive APs and reduce the transmit power. In the previous example, if the transmit power was 0 dBm (1 mW) instead of 30 dBm (1 W), the distance $d$ would be 21 m, which results in a much smaller positioning error. Another possibility is to exploit the *clear channel assessment* (CCA) thresholds used in the case of Wi-Fi communications for supporting carrier sense multiple access (Mhatre, Thompson,

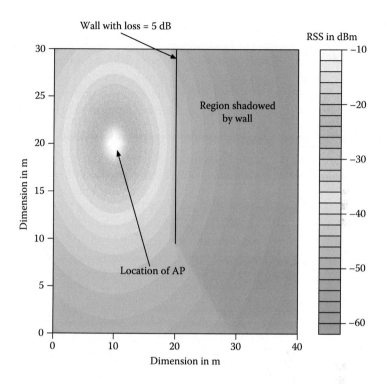

**FIGURE 3.2**
RSS from an AP and illustration of the shadowing problem.

Papagiannaki, & Baccelli, 2007). The CCA threshold allows devices to ignore transmissions that have RSS values below a certain threshold. Setting these values to be higher than −80 dBm may reduce the range of a Wi-Fi AP as necessary.

*Subarea localization:* We can consider as an improvement to proximity-based positioning the idea of *subarea* localization that was proposed in Aksu and Krishnamurthy (2010). This work considers using information about which APs can be "heard" by a mobile device rather than the device simply considering the proximity to an AP. In most indoor areas of today, it is possible to "see" many APs at a given position, but the set of these APs changes with the position in the building. For example, let us suppose that there are four APs located at the corners of a square geographical area. If the coverage of these APs can be tuned and assumed to be circular, it is possible to create 13 unique subareas that are covered by *unique* subsets of these APs. Based on this information, the mobile device can be located in a subarea that is covered by *exactly* a given set of APs. This subarea can be small or large depending on the placement and coverage of APs impacting the accuracy and precision of positioning. If we denote the existence or lack thereof of an AP in the visibility set by a 0 or a 1, we can see that there are 16 unique fingerprints that are possible in this example, namely, [0 0 0 0] through [1 1 1 1], but it is not possible for certain fingerprints to exist, setting the maximum number of unique fingerprints at 13 (see Aksu and Krishnamurthy, 2010, for more details). This number could be smaller, for instance, if all four APs cover the square area, the only fingerprint is [1 1 1 1], which indicates that the mobile device is *somewhere* in the square area. Multifloor subarea localization requires a different algorithm to identify the floor on which a device with an unknown position is located (Korkmaz, 2011).

Subarea localization used in conjunction with CCA thresholds could provide a simple yet fairly accurate positioning approach for indoor areas. However, the work in Aksu and Krishnamurthy (2010) and Korkmaz (2011) is theoretical and based on simulations. There are no real implementations that have evaluated the performance of subarea localization.

*Location fingerprinting:* Let us denote the vector of APs that are visible at a certain location by $\rho$. In the case of subarea localization, $\rho$ is a vector with elements that are either zero or one. Fine-grained location fingerprinting extends this idea as follows: instead of recording the visibility or not of an AP, the average RSS values from that AP or, in some cases, the distribution of the RSS samples from an AP are recorded. This is called the *location fingerprint* $\rho$. The determination of the position of the mobile device using location fingerprinting typically has two phases. In the *offline phase*, a laborious survey of the area to be covered by the positioning system is performed to collect the fingerprints at various positions to populate a database. Sometimes this is also referred to as the training set. In the *online phase*, the unknown position of a mobile device is estimated. The typical approach is to compare the measured vector of RSS by a mobile device with the entries of fingerprints in a database to estimate and determine the *closest* fingerprint. The closeness is typically estimated using a similarity metric such as the Euclidean distance (Bahl & Padmanabhan, 2000). The position associated with this closest fingerprint is returned as the estimated location. In the seminal work on Wi-Fi location fingerprinting in (Bahl & Padmanabhan, 2000), the authors determined the median error with Wi-Fi location fingerprinting to vary between 3 m and 6 m depending on the number of grid points employed with three APs visible (i.e., the dimension of $\rho$ is 3) using a deterministic nearest-neighbor scheme. With a similar deterministic algorithm, the cumulative distribution function of errors in Swangmuang and Krishnamurthy (2008) indicated that the probability that the positioning error is smaller than 4 m is 0.9 in an office area with 25 grid points and three visible APs. In Youssef, Agrawala, and Shankar (2003), experiments were conducted in an office-like building that measured roughly 225 feet by 85 feet. A total of 110 grid points were used to create the fingerprint database. Most of the fingerprints were collected along long corridors. The cumulative distribution function of positioning errors indicates that the errors are smaller than 7 feet (about 2.1 m) with a probability of 0.9.

A majority of Wi-Fi-based positioning approaches employ location fingerprinting. The major challenge with location fingerprinting is the labor-intensive nature of collecting location fingerprints. Recently, crowd-sourced collection of location fingerprints (e.g., see Park et al., 2010) has been suggested to reduce the labor effort for location fingerprinting. For a comprehensive discussion of Wi-Fi fingerprinting, the reader is referred to Krishnamurthy (2013).

### 3.2.2 Cellular-Telephone-Based Positioning

With the Federal Communications Commission's Enhanced-911 (E-911) mandate, cellular telephony has included positioning as part of its standards for more than a decade now. The E-911 mandate required that mobile phones be located with an accuracy of at least 50 m for 67% of the emergency calls. GPS has been used for localizing E-911 calls, although cell-phone-network-based localization is used in conjunction for quicker fixes of the position. The approaches using the cell phone network for localization typically employ TOA/TDOA as the positioning method, with technology-specific protocols. For example, in 3G UMTS, the protocol is called Observed Time Difference of Arrival–Idle Period on DownLink (OTDOA-IPDL) (Porcino, 2001), while it is called Advanced Forward Link Trilateration (A-FLT) in 2G CDMA-based cellular networks (Nissani & Shperling, 2000).

As already discussed, an accuracy of 50 m is not sufficient for indoor areas. In the case of positioning with Wi-Fi, it is possible to obtain accuracies on the order of 2 m to 6 m depending on the environment and algorithm with location fingerprinting. Thus, it appears that location fingerprinting is an appropriate method for positioning that may perhaps be used with signals from cell phone towers as well. This very idea was examined in detail in Otsason, Varshavsky, LaMarca, and de Lara (2005) and Varshavsky, de Lara, Hightower, LaMarca, and Otsason (2007) with the use of signals from the Global System of Mobile Communications (GSM) technology. GSM is a 2G cellular telephone technology (Pahlavan and Krishnamurthy, 2013) that had around 4 billion connections in 2009. While GSM is being replaced by 3G and 4G services worldwide, it still has perhaps the largest global subscriber base. In GSM, it is required by mobile stations to monitor up to six neighboring base stations in addition to the one where a telephone call may be placed. The work in Varshavsky et al. (2007) uses signals from up to 29 additional base stations to create a location fingerprint. This drastically increases the dimensionality of the fingerprint, thereby improving accuracy. The experimental work in Varshavsky et al. (2007) considered the use of GSM location fingerprints in multifloor buildings in three different cities. The accuracy and precision were comparable or better than that with IEEE 802.11-based fingerprinting (median location errors were around 2 m to 3 m compared to 2 m to 5 m for Wi-Fi). Floors were recognized correctly about 60% of the time, and the location was within two floors about 98% of the time.

The reference signal used for fingerprinting-based positioning becomes important since RSS measurements are based on such reference signals. In the case of Wi-Fi, a medium access control management frame, called the *beacon* frame, is used for RSS measurements. In the case of GSM, the *broadcast control channel* is used for RSS measurements. If similar approaches are used with 3G UMTS or 4G LTE cellular services, the reference signals will change.

### 3.2.3 Bluetooth-Based Positioning

Bluetooth (Pahlavan & Krishnamurthy, 2013) is often classified as a wireless personal area networking technology, where the *personal* network occupies a limited space around a human being. It is mostly used for connections between the various devices that an individual may be using, such as cameras, smartphones, laptops, and computers. Consequently, the transmit power and range of Bluetooth are much smaller than the range of Wi-Fi (typically on the order of 10 m).* Bluetooth has been considered for indoor positioning in works such as those reported in Fischer, Dietrich, and Winkler (2004), Anastasi et al. (2003), and Bekkelien (2012).

The work in Anastasi et al. (2003) uses *proximity-based* positioning with Bluetooth. Fixed workstations equipped with a Bluetooth interface keep track of mobile devices (with Bluetooth interfaces) that are within their coverage area. The Bluetooth device address is used to track devices as they move (e.g., to determine whether they have left the coverage area). A centralized server is used to monitor all of the devices in the indoor area of interest. Apple's iBeacon technology (see *iOS: Understanding iBeacon*, http://support. apple.com/kb/HT6048) also uses proximity of an iOS device using Bluetooth to determine its approximate location.

In Fischer et al. (2004), the time of arrival of a Bluetooth signal is used for estimating the range or distance of a mobile device from known fixed receivers. The fixed receivers are modified with additional hardware to determine the time of arrival up to nanoseconds. To avoid the problem of time synchronization across devices, a differential time difference of arrival (DTDOA) method is used for positioning. The "Echo" request and response messages are used for computing the time of arrival of signals from a mobile device. While this prototype showed promise, the results described in Fischer et al. indicate that TOA measurements are severely impacted by multipath and movement of people and objects in the Bluetooth environment. Furthermore, the experiments were very simple and used only two fixed receivers, moving the mobile device in between them.

Location fingerprinting with Bluetooth headsets was employed in Bekkelien (2012) for positioning. Various fingerprint similarity metrics were

---

* Bluetooth has several device classes whose transmit powers and capabilities vary. Thus the range can be as small as 5 m or as large as 100 m. See the official Bluetooth website, http://www.bluetooth.com/Pages/Bluetooth-Home.aspx, for details.

used, with RSSI-based location fingerprints. The accuracy of the position estimates was around 2 m. In terms of precision, the location errors were smaller than 6 m, 95% of the time with all of the algorithms.

### 3.2.4 RFID-Based Positioning

RFID technology has gained significant importance in recent years as a replacement for bar codes and for tracking inventory and objects in a variety of applications (Garfinkel & Rosenberg, 2006). In the simplest case, a *passive* RFID tag is like a bar code that can be read by a more powerful *reader*. The way a passive tag works is that a signal is sent by the reader, which is reflected back with the tag-specific modulation (e.g., its identification). The passive tag needs no battery power, making it an inexpensive device for deployment. Active RFID tags have their own source of power and can transmit signals that can be detected by readers. The disadvantage of using RFID tags for positioning is that this technology is not typically built into smartphones, which are regularly being used for navigation today. This is in contrast to the other three technologies (Wi-Fi, Bluetooth, and cellular) that are present in most smartphones of today.

Like Wi-Fi, cellular, and Bluetooth technologies, RFID technology can be employed for positioning in various ways such as proximity-based positioning or fingerprinting. In the case of proximity-based positioning, the range of a passive (or active) tag determines the location accuracy. This can be very small or on the same order as Bluetooth depending on the nature of the tag. In the case of location fingerprinting, it is possible to use the RSS measurements from known RFID tags for determining the location. However, a different twist is possible if inexpensive passive RFID tags are widely deployed. It may be possible to estimate the position of a reader based on the number and identities of the responding RFID tags, which itself can be considered as a location fingerprint. The reader is referred to Ward, Jones, and Hopper (1997) and Hazas and Hopper (2006) for example work on indoor positioning with RFIDs.

### 3.2.5 Other RF Positioning Technologies

The accuracy of RF positioning, especially using time of arrival of signals, is impacted by the bandwidth and design of the RF signals. As mentioned previously, the signals used by Wi-Fi, GSM, and Bluetooth are not designed for positioning purposes. Thus they perform poorly with TOA and other traditional positioning algorithms.

*Ultrawideband* or UWB signals have extremely large signal bandwidths (on the order of GHz) that can be used for fine-grained ranging. The disadvantage of UWB is that the transmit powers are very small because of regulatory restrictions. They have also not seen commercial success. A detailed discussion of UWB is beyond the scope of this chapter, but the reader is referred to

Pahlavan and Krishnamurthy (2013) and the references therein for a treatment of UWB and its use in range estimation for localization.

---

## 3.3 Non-RF-Based Positioning

While RF-based indoor positioning appears to be the best solution because of its almost ubiquitous presence in smartphones, non-RF-based positioning schemes have also received attention over the past decade because of the limitations of accuracy of RF-based positioning schemes. They have also been investigated as potential alternatives that may find their use in niche applications. In this section, we briefly discuss some of these approaches.

### 3.3.1 Acoustic Positioning

As early as 1997, the *Active Bat* project (Ward et al., 1997) demonstrated the use of ultrasound as an approach for locating objects in indoor areas. In this work, the time of arrival of ultrasonic signals was used to determine the range of objects. The accuracy of this system was on the order of a few centimeters. Clearly, this is two or three orders of magnitude better than the accuracy obtained with typical RF technologies such as Wi-Fi or Bluetooth, which provide accuracies on the order of several meters. The disadvantages of ultrasound or any acoustic technique are twofold. First, ultrasound or sound cannot penetrate walls and typically need line of sight. Second, like RFID technology, most smartphones used for navigation are not equipped with ultrasound technology.

Irrespective of these disadvantages, in Hazas and Hopper (2006), other disadvantages of the work in Ward et al. (1997) were enumerated. These include the inability to localize more than one object at a time, the need for RF to identify the object, and the susceptibility to ultrasonic noise (e.g., the clinking of a pencil). To overcome these problems, Hazas and Hopper (2006) proposed the use of spread-spectrum for ultrasonic ranging. This system employs Gold codes, which are also employed by GPS for ranging. Accuracies on the order of a centimeter or better in three dimensions were observed. In Sertatıl, Altınkaya, and Raoof (2012), the same 511 chip Gold code used in Hazas and Hopper (2006) was used with acoustic signals (not ultrasonic signals) for a location accuracy of 2 cm, 99% of the time, using inexpensive off-the-shelf speakers and microphones. We note here that the speakers are the transmitters whose locations are known, and the microphone is the device whose position needs to be estimated.

### 3.3.2 Dead Reckoning or Inertial Navigation

Dead reckoning refers to a *relative* positioning approach that starts from a known location and tries to take into account the movement of an object in terms of speed, distance, and direction to estimate the new location. As in the case of other positioning schemes, it is possible that errors can be introduced into dead reckoning schemes as well, resulting in large position errors. Smartphones are now equipped with accelerometers that can be used to determine how many steps a person may have taken from a previous location. The availability of sensors that can detect the magnetic direction also facilitates dead reckoning. We note here that this approach needs a starting known location, which can be problematic in some situations.

As one example of work done in this area, pedestrian dead reckoning was considered in Beauregard and Haas (2006) with the use of walking speed and displacement estimation in terms of the number of steps taken by a person. The number of steps taken was sensed by an accelerometer worn by the individual during the walk from the known location to an unknown location. A neural network was used to train a model for estimating the number of steps based on the accelerometer readings. The accuracy was within 10 m after a 1 km walk. As another example, a foot-mounted magnetic sensor for inertial navigation was reported in Bird and Arden (2011), which uses a MEMS inertial measurement unit for determining the relative displacement of soldiers from a known location. Experiments showed that the estimated track stayed within 2 m of the real track of a person walking in an entire in-building run.

### 3.3.3 Signage and Maps

The most widely available approach for human navigation is the use of signs and maps that need no additional infrastructure such as power, devices, spectrum, processing, or databases. Most malls have mall directories, which indicate the location of the directory in a map, which also shows other points of interest such as stores (classified by category), restrooms, elevators, escalators, and different floors and annexes where they exist. In most buildings, a building directory provides information on how to reach specific individuals or businesses (e.g., Prof. Karimi is on the seventh floor, and on the seventh floor, an arrow points to a set of room numbers). Recently, there are other types of "signs" and "maps" that are being used for positioning and navigation. We discuss two such approaches below.

*Smartphone cameras:* The simplest way to use smartphone cameras is to use "marker" signs in areas. These are similar to the quick response or QR codes (a matrix-like bar code) that can be scanned by an application in a smartphone's camera. Based on the unique identity of these markers, the location of the phone can be determined and displayed by the application. Such an approach is described in Mulloni, Wagner, Schmalstieg, and Barakonyi (2009). A more sophisticated approach described in Hile and Borriello (2008) tries to match features of a building (such as corridors and intersections) scanned by a phone's camera with those in a database of the building's floor plan. When a match is found, the corresponding map and location is displayed on the user's phone.

*Acoustic BG:* Instead of matching the building features, the work in Tarzia, Dinda, Dick, and Memik (2011) examined the possibility of matching what they called the *acoustic background spectrum* of a room. The granularity is that of a room in this case, and the premise is that each room has its own acoustic signature that depends on geometry and furnishings that can be compared to one captured in a database to determine one's location. The experiments of Tarzia et al. (2011) provided a correct match 69% of the time in a test composed of 33 rooms. The acoustic background spectrum signature is about 1.3 kB per room and needs 10 s to be acquired by a smartphone's microphone.

### 3.3.4 Hybrid Schemes

Considering the range of approaches that we have discussed so far, an obvious question that comes to mind is whether there are advantages of combining and fusing the positioning information from multiple technologies to improve accuracy, precision, and latency in significant ways. Unfortunately, indoor positioning is still in its infancy, and there are no systematic evaluations of the benefits or trade-offs of combining or fusing various approaches. One potential drawback is the consumption of battery resources by multiple technologies if they are used simultaneously for positioning purposes. Another drawback is the handling of conflicting positions or locations.

Nevertheless, we can see some examples of hybrid schemes that try to enhance the performance of positioning in indoor areas. Some systems that we previously mentioned, such as the Active Bat (Ward et al., 1997), already employ both RF and ultrasound technologies. The use of WLAN fingerprinting with RFID tags was explored in Spinella, Iera, and Molinaro (2010). This needs the mobile device at an unknown location to have both a Wi-Fi card and an RF reader built into it, which may increase its bulk and reduce its battery life. Combining RF/ultrasound with inertial navigation was examined in Popa, Ansari, Riihijarvi, and Mahonen (2008). The work shows that inertial navigation can help locate a mobile in areas where RF/ultrasound does not provide enough coverage. Finally, in Holm (2009), a prototype combination of ultrasound and RFID was demonstrated for indoor localization.

## 3.4 Discussion

There is a plethora of potential technologies for indoor positioning, none of which have made deep inroads toward taking over the market. There are several reasons for this. The applications that need positioning information and their requirements vary significantly. In smaller indoor areas, signs are sufficient for human beings to navigate their way to their destinations. Larger areas such as malls and airports have specific needs. For example, an individual in an airport has to go through security checks and proceed to his or her gate, a process that is well structured. Airports are architecturally designed to facilitate this movement. Resources such as restrooms are dispersed adequately to eliminate the need to search for directions. It is also cumbersome for a human being to keep looking at a map while walking to position himself in his environment. However, there is significant interest in this for emergencies and disasters, for general situational awareness, and, over time, for locating other resources in indoor areas may need accurate positioning and maps.

One aspect of these technologies that has not been discussed in this chapter is the latency or time to positional fix. In Anastasi et al. (2003), the proximity-based localization with Bluetooth was estimated to take more than 1 s. Location fingerprinting with Wi-Fi requires scanning for many APs and frequencies that will likely incur significant delay. It may be the case that dead reckoning will be helpful in filling the gaps for navigation during such delays.

## 3.5 Summary

This chapter provides an overview of technologies that can be used for indoor positioning. While GPS dominates for outdoor positioning and navigation, there is no such single standard technology of choice (yet) for indoor areas. Research works have analyzed and evaluated several technologies as alternatives for positioning in indoor areas.

Such technologies are roughly classified into RF-based and non-RF-based approaches in this chapter. The most popular RF-based approach, namely, Wi-Fi, is used to discuss the methodologies used for positioning, such as proximity to a known device and location fingerprinting. As discussed later with other technologies such as Bluetooth and RFID, these two methodologies dominate indoor positioning, because of the problems with multipath and radio propagation vagaries in indoor areas. The time of arrival of signals is employed with large bandwidth UWB signals and with ultrasound or acoustic signals. However, they are not very popular commercially, because of their sparse presence in smartphones.

Recently, the use of inertial navigation either by itself or in combination with RF-based positioning has found some interest, along with nontraditional approaches such as scanning markers with phone cameras or matching the acoustic signature of a room with those stored in a database. Hybrid approaches that combine technologies are also being explored. The field is still in its infancy, and it is not clear how the evolution of these technologies will occur toward a standard for indoor positioning.

# References

Aksu, A., & Krishnamurthy, P. (2010, October). *Sub-area localization: A simple calibration free approach*. Proceedings of the 13th International Symposium on Modeling Analysis and Simulation of Wireless and Mobile Systems, MSWiM 2010, Bodrum, Turkey.

Anastasi, G., Bandelloni, R., Conti, M., Delmastro, F., Gregori, E., & Mainetto, G. (2003). Experimenting an indoor Bluetooth-based positioning service. In *Proceedings of the 23rd international conference on distributed computing systems (ICDCS) workshops*. New York, NY: IEEE.

Bahl, P., & Padmanabhan, V. N. (2000). Radar: An in-building RF based user location and tracking system. In *INFOCOM 2000* (pp. 775–784). New York, NY: IEEE.

Beauregard, S., & Haas, H. (2006). Pedestrian dead reckoning: A basis for personal positioning. In *Proceedings of the 3rd workshop on positioning, navigation, and communication*. Germany: Shaker Verlag.

Bekkelien, A. (2012). *Bluetooth indoor positioning* (Unpublished master's thesis). University of Geneva, Switzerland.

Bird, J., & Arden, D. (2011). Indoor navigation with foot-mounted strapdown inertial navigation and magnetic sensors. *IEEE Wireless Communications, 18*(2), 28–35.

Deak, G., Curran, K., & Condell, J. (2012). A survey of active and passive indoor localisation systems. *Elsevier Computer Communications, 35*, 1939–1954.

Fischer, G., Dietrich, B., & Winkler, F. (2004). Bluetooth indoor localization system. In *Proceedings of the 1st workshop on positioning, navigation, and communication*. Germany: Shaker Verlag.

Garfinkel, S., & Rosenberg, B. (Eds.). (2006). RFID: Applications, security, and privacy. Boston, MA: Addison-Wesley.

Gu, Y., Lo, A., & Niemegeers, I. (2009). A survey of indoor positioning systems for wireless personal networks. *IEEE Communications Surveys and Tutorials, 11*(1), 13-32.

Hazas, M., & Hopper, A. (2006). Broadband ultrasonic location systems for improved indoor positioning. *IEEE Transactions on Mobile Computing, 5*(5), 536–547.

Hile, H., & Borriello, G. (2008). Positioning and orientation in indoor environments using camera phones. *IEEE Computer Graphics and Applications, 28*(4), 32–39.

Holm, S. (2009). Hybrid ultrasound-RFID indoor positioning: Combining the best of both worlds. *RFID, 2009 IEEE International Conference on*. New York, NY: IEEE.

Korkmaz, A. (2011). *On proximity based sub-area localization* (Unpublished doctoral dissertation). University of Pittsburgh, Pennsylvania.

Krishnamurthy, P. (2013). WiFi location fingerprinting. In H. Karimi (Ed.), *Advanced location-based technologies and services*. Boca Raton, FL: CRC Press.

Mhatre, V. P., Thompson, P., Papagiannaki, K., & Baccelli, F. (2007). Interference mitigation through power control in high density 802.11 WLANs. In *INFOCOM 2007: 26th IEEE international conference on computer communications*. New York, NY: IEEE.

Mulloni, A., Wagner, D., Schmalstieg, D., & Barakonyi, I. (2009). Indoor positioning and navigation with camera phones. *IEEE Pervasive Computing, 8*(2), 22–31.

Nissani, D. N., & Shperling, I. (2000). Cellular CDMA (IS-95) location, A-FLT proof-of-concept interim results. In *The 21st IEEE convention of the electrical and electronic engineers in Israel, 2000* (pp. 179–182). New York, NY: IEEE.

Otsason, V., Varshavsky, A., LaMarca, A., & de Lara, E. (2005). Accurate GSM indoor localization. In M. Beigl, S. Intille, J. Rekimoto, & H. Tokuda (Eds.), *UbiComp 2005: Ubiquitous computing* (Vol. 3660, pp 141–158). Berlin, Heidelberg, Germany: Springer-Verlag.

Pahlavan, K., & Krishnamurthy, P. (2013). *Principles of wireless access and localization.* Hoboken, NJ: John Wiley and Sons.

Park, J.-G., Charrow, B., Curtis, D., Battat, J., Minkov, E., Hicks, J., Teller, S., & Ledlie, J. (2010). Growing an organic indoor location system. In *MobiSys '10: Proceedings of the 8th international conference on mobile systems, applications, and services.* New York, NY: ACM.

Perahia, E., & Stacey, R. (2008). *Next generation wireless LANs.* Cambridge, UK: Cambridge University Press.

Popa, M., Ansari, J., Riihijarvi, J., & Mahonen, P. (2008). Combining cricket system and inertial navigation for indoor human tracking. In *Wireless communications and networking conference.* New York, NY: IEEE.

Porcino, D. (2001). Performance of a OTDOA-IPDL positioning receiver for 3gpp-fdd mode. In *Second international conference on 3G mobile communication technologies, 2001* (pp. 221–225). IET.

Sertatıl, C., Altınkaya, M. A., & Raoof, K. (2012). A novel acoustic indoor localization system employing CDMA. *Digital Signal Processing*, 22(3), 506–517.

Spinella, S. C., Iera, A., & Molinaro, A. (2010). On potentials and limitations of a hybrid WLAN-RFID indoor positioning technique. *International Journal of Navigation and Observation, 2010*, 1–11.

Swangmuang, N., & Krishnamurthy, P. (2008). Location fingerprint analyses toward efficient indoor positioning. In *PerCom'08* (pp. 100–109). New York, NY: IEEE.

Tarzia, S. P., Dinda, P. A., Dick, R. P., & Memik, G. (2011). Indoor localization without infrastructure using the acoustic background spectrum. In *MobiSys '11: Proceedings of the 9th international conference on mobile systems, applications, and services.* New York, NY: ACM.

Varshavsky, A., de Lara, E., Hightower, J., LaMarca, A., & Otsason, V. (2007). GSM indoor localization. *Pervasive and Mobile Computing*, 3(6), 698–720.

Ward, A., Jones, A., & Hopper, A. (1997). A new location technique for the active office. *IEEE Personal Communications*, 4(5), 42–47.

Youssef, M. A., Agrawala, A., & Shankar, A. U. (2003). WLAN location determination via clustering and probability distributions. In *Proceedings of the first IEEE international conference on pervasive computing and communications, 2003 (PerCom 2003).* (pp. 23–26). New York, NY: IEEE.

# 4

# Magnetic Indoor Local Positioning System

Jörg Blankenbach
Abdelmoumen Norrdine

## CONTENTS

*Abstract:* In past years, many technologies have been evaluated for positioning and navigation tasks inside buildings, resulting in various indoor positioning systems. Many systems are infrastructure based using electromagnetic waves (e.g., radio waves) or ultrasound for signal propagation between fixed transmitters and mobile receivers (or vice versa). The main drawback of these systems is signal propagation error due to attenuation, shadowing, multipath effects, or signal delay inside buildings caused by building objects (e.g., walls, ceilings, and furniture). Even if some technologies are more robust against these effects than others (e.g., ultrawide band), it is impossible to suppress signal propagation errors completely. However, contrary to electromagnetic waves, magnetic signals are able to pass through any building material without propagation errors, even in non-line-of-sight (NLoS) scenarios. In this chapter, an alternative indoor positioning system based on artificially generated magnetic fields, which is able to cope with all these effects and enables a reliable localization even under harsh conditions, is presented. By using the observed magnetic signals from at least three different and spatially distributed coils, the 3D position of a mobile station can be determined by trilateration. Although the system is still under development, 3D positioning accuracies, by using three coils and an industrial magnetic field sensor, in the range of centimeters to a few decimeters, are already achievable in a laboratory scale.

## 4.1 Introduction

The majority of electromagnetic- or acoustic-waves-based indoor positioning systems face many difficulties in indoor environments because of measurement errors caused by fading, shadowing, multipath effects, or signal propagation delay inside buildings. Because magnetic fields are able to penetrate walls, building materials, or other objects without attenuation, a direct current magnetic signal-based indoor local positioning system (MILPS) was developed. In this chapter, some related work, the principle of the MILPS, and the underlying magnetic field theory are presented in Section 4.2. Under consideration of simulation results, a demonstrator, consisting of three coils and a triaxial magnetometer, is also shown in Section 4.2. In Section 4.3, we describe a differential measurement principle and noise suppression algorithms for the elimination of interference fields (e.g., the earth's magnetic field or electrical disturbances). Several range measurements in real indoor scenarios were carried out. The results are discussed in Section 4.4. In Section 4.5, 2D and 3D positioning are examined, while the last section gives a conclusion and a short outlook on further research.

## 4.2 MILPS: Magnetic Indoor Local Positioning System

### 4.2.1 Magnetic Field-Based Localization

When magnetic fields are used for indoor localization, a distinction must be made between systems based on natural magnetic fields (geomagnetism, magnets) and those based on artificially generated magnetic fields.

*Geomagnetism:* The earth's magnetic field (geomagnetic field) is often used for heading estimation in navigation applications. Indoor environments, however, contain various components, such as ferrous structural materials, furniture, and electrical currents, which disturb the earth's natural magnetic field. The resulting variations in the magnetic field in indoor environments can be used as a fingerprint to identify a user's position and possibly his or her orientation (B. Li, Gallagher, Dempster, & Rizos, 2012). A similar approach was taken by Subbu, Gozick, and Dantu (2013), who collected magnetic signatures within a building by using a magnetic field sensor integrated in a smartphone. Position estimation was performed based on this signatures classification.

*Permanent magnets:* Another approach that uses the magnetic flux density for positioning is through magnetic fields created by permanent magnets. A system consisting of magnetic field sensors placed at known locations in a 50 cm × 50 cm × 50 cm cube and a mobile permanent magnet is described in Mao Li et al. (2009). The achievable localization accuracy in the small-scale measurement volume is indicated by a few millimeters.

*Artificially generated magnetic fields:* Object tracking by means of artificially generated magnetic fields has been examined in recent decades (Kuipers, 1975; Prigge & How, 2004; Raab, Blood, Steiner, & Jones, 1979). Two commercial systems are already on the market.* However, a majority of magnetic-fields-based tracking systems are designed for motion tracking and virtual reality. These systems are mainly used for artistic, industrial, and biomedical applications in specially equipped laboratory environments (Callmer, Skoglund, & Gustafsson, 2010; Hu, Meng, Mandal, & Wang, 2006). Magnetic field creation is usually accomplished with the use of concentric coils and is limited to small measurement volumes only (typically with a radius less than 3 m) (Blood, 1990; Dong et al., 2004; Paperno, Sasada, & Leonovich, 2001; Raab, 1982). Systems that use sinusoidal magnetic fields are described in Kuipers (1975) and Raab et al. (1979). In this case, the magnetic fields have to first be filtered by frequency on the receiver side. Blood (1990) and Anderson (1995) described systems using pulsed direct current fields. Here are the fields which are generated sequentially. Prigge (2004) presented an experimental system that utilizes eight small diameter coils giving a limited size coverage area of 4 m × 4 m. To distinguish the signals from the different coils, the researchers used a code division multiple access approach (CDMA). Callmer et al. (2010) explored the usage of triaxial magnetometers and a vessel with known magnetic dipole to localize the sensors in underwater environments. Hu et al. (2006) presented a magnetic localization and orientation system for medical diagnoses and treatments to wirelessly track an object through the human gastrointestinal tract.

---

* See Ascension Corporation, http://www.ascension-tech.com; Polhemus Corporation, http://www.polhemus.com.

### 4.2.2 System Architecture and Function Principle of the MILPS

The objective of the MILPS is to provide a reliable and accurate indoor positioning system that covers an entire building with a minimum of infrastructure and complexity.

To enable position estimation, similar to other infrastructure-based systems, reference stations (RSi) are placed inside or outside the building. The RSi in this case are represented by electrical coils (see Figure 4.1) for artificial magnetic field generation.

For the position determination, a mobile station (MS) is equipped with a magnetic field sensor (magnetometer). By measuring the field components of multiple (at least three) coils, we can determine the distances between the RSi and the MS. Based on the coordinates $(X_i, Y_i, Z_i)$ of the RSi in the building reference system, the unknown 3D coordinates of MS $(X_{MS}, Y_{MS}, Z_{MS})$ can be estimated by applying the trilateration principle.

### 4.2.3 Magnetic Field of a Coil

The magnetic field is a physical state of space described as a vectorial physical quantity. The presence of magnetic fields can be ascertained by the existence of forces on magnets, magnetized matter, or moving charge carriers (Lorentz forces). In the following, only the induction $B$ is considered and, for reasons of simplicity, is called "magnetic field." Theoretically, there are

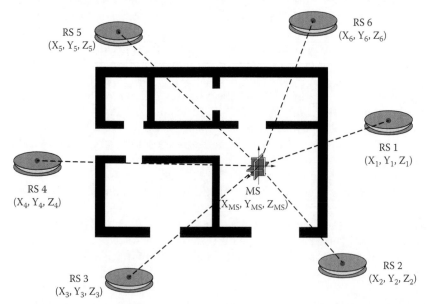

**FIGURE 4.1**
The MILPS system architecture with reference stations (RS 1–RS 6) and a mobile station (MS).

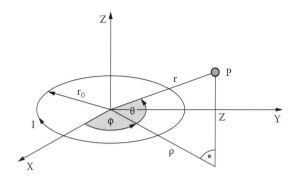

**FIGURE 4.2**
Where $r_0$ is the coil radius, $\theta$ is the elevation angle, $r$ is the distance between the position P and the coil's center.

an infinite number of ways to produce magnetic fields. In practice, current coils are very suitable for that purpose because of the simplicity of the coil's structure and the accordance of the measured and the theoretically calculated magnetic field. For the description of the coil's magnetic field, we first discuss the field of a circular current loop.

By using spherical coordinates and assuming that the position P is at a distance $r$ that is several times the loop radius $r_0$, the three components $B_r$, $B_\theta$, and $B_\varphi$ of the magnetic field vector (see Figure 4.2) can be expressed by using the following equations (Lehner, 2006):

$$B_r = \frac{\mu_0 NIF}{2\pi r^3}\sin(\theta)$$

$$B_\varphi = 0 \tag{1}$$

$$B_\theta = \frac{\mu_0 NIF}{4\pi r^3}\cos(\theta)$$

The product of the number of turns of wire $N$, the loop's current intensity $I$, and the coil's area $F$ is called the magnetic dipole moment $p_m = NIF$. The total magnetic field $B$ at the point P $(r, \theta, \varphi)$ can be calculated from Equation 1 using the following relation:

$$B = \frac{\mu_0 NIF}{4\pi r^3}\sqrt{1 + 3\sin^2\theta}. \tag{2}$$

### 4.2.4 Experimental System

*Simulation:* To get an estimate of the field strength in the coil's coverage area, software simulations with varying coil parameters using the equation for field strength calculation (Equation 2) were performed. For simplicity, it was assumed that the coil and the unknown point (magnetic field sensor) were in the same horizontal plane ($\theta = 0$). Thus, the distance $r$ can be resolved from Equation 2, as seen in the following equation:

$$r = \sqrt[3]{\frac{\mu_0 N I F}{4 \pi B}}. \tag{3}$$

Figure 4.3 (top) depicts a simulation result for different current strengths $I = \{10 \text{ A}, 14 \text{ A}, 18 \text{ A}\}$ using the constant coil parameters $N = 140$, $r_0 = 0.25$ m.

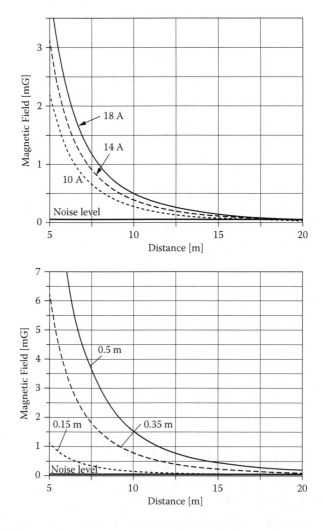

**FIGURE 4.3**
Calculated magnetic field strength at different current strengths (top, $r_0 = 0.25$ m, $N = 140$) and at a different coil radius (bottom, $I = 14$ A, $N = 140$).

From the curves (see Figure 4.3) it can be inferred that the coil's magnetic field decreases very rapidly with the distance *r*. However, distances up to 18 m are still resolvable at a noise level of about 70 μG.

Because of the flat curve at large distances, the determination of the distance *r* from the observed magnetic field becomes inaccurate. An estimate for accuracy of the distance determination, depending on the accuracy of the observed magnetic field strength *B*, can be calculated with the law of variance (error) propagation assuming that there are no systematic errors. Figure 4.4[*] depicts the standard deviation as a function of the distance *r* for different standard deviations $\sigma_B$ of the magnetic field. Accuracy in the millimeter and centimeter range can be achieved at short distances (up to 4 m) and medium distances (4 m–7 m). For larger distances, the accuracy of the distance determination is reduced significantly. For example, at a distance of 20 m, the standard deviation $\sigma_r$ is about 1.4 m if $\sigma_B = 10$ μG. This illustrates that in addition to the coil, the choice of the magnetic field sensor (see Table 4.1) plays a crucial role for the development of the MILPS.

With respect to realization and implementation of an accurate positioning system, it is necessary to use sensors that are available on an industrial scale. These are magneto resistive sensors, Hall sensors, or fluxgate sensors. In Table 4.1 the characteristics of different sensors are compared.

*Prototype:* For the realization of an experimental system, the previously described simulation results were considered. The desired maximal range

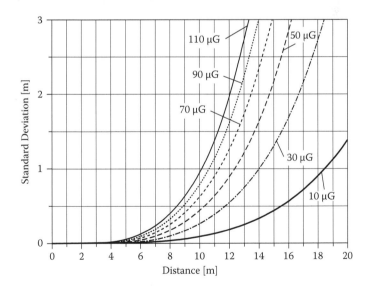

**FIGURE 4.4**
Simulated distance precision for different standard deviations of the magnetic field.

---

[*] Where $r_0$ is the coil radius, $\theta$ is the elevation angle, *r* is the distance between the position P and the coil's center.

**TABLE 4.1**

Characteristic Values of Some Commercial Magnetic Field Sensors

| Physical Effect | Device | Measurement Range | Resolution | Bandwidth |
|---|---|---|---|---|
| AMR | HMC1001 | ±2 G | 27 µG | 0–5 MHz |
| Three-axis FG | FGM-3h | ±0,5 G | — | 20 kHz |
| SDT | NVE | ±2 G | 10 µG | 125 Hz |
| MI | MicroMag3 | ±11 G | 150 µG | 1 kHz |
| GMR | AAL002 | ±30 G | 100 µG | >1 MHz |
| Hall effect | MFS-3A | ±73 G | 100 mG | <100 kHz |

*Note:* AMR = anisotrope magneto resistive, FG = fluxgate, SDT = spin dependent tunneling, MI = magneto inductive, GMR = giant magneto resistance.

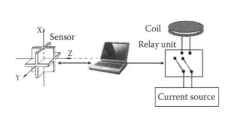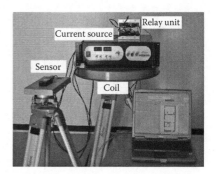

**FIGURE 4.5**
Experimental system with a 3-axis magnetic field sensor, coil, relay unit, power supply and measuring PC.

of 15 m to 20 m between sensor and coil on the one hand (see Figure 4.3) and some pragmatic considerations regarding the dimensions, the weight, and the maximum current of the coils on the other hand led to a coil consisting of $N = 140$ turns of wire wrapped on $r_0 = 0,5$ m diameter core. The electrical current is provided by a low-cost power supply, and a relay unit enables the switching of the currents. Magnetic field sensing is accomplished using three Honeywell magneto resistive transducers aligned in orthogonal directions, allowing the capturing of the three vector magnetic field components.[*] The sensor's measurement range is between ±2 G, and the resolution is 67 µG, which corresponds to the noise level chosen for the simulation.

Figure 4.5 shows the setup of the experimental system. The relay unit, the coil's power source, and the reading of the sensor data (maximal sampling time $T_s = 6.5$ ms) are controlled by a software application running on a laptop computer.

---

[*] See Honeywell, http://www.Honeywell.com/.

## 4.3 Measurement Principle and Signal Processing

For the application of the experimental system inside of buildings, it has to be borne in mind that further magnetic fields (e.g., the earth's magnetic field, the field from the electricity in buildings) interfere with the coil's field (see Figure 4.6). Thus, a differential measuring principle for eliminating the interference fields has been applied. Therefore, the coil's current direction is periodically switched. Figure 4.7 depicts the raw data of a measurement example

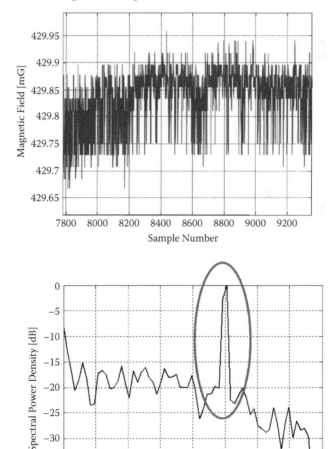

**FIGURE 4.6**
Measured magnetic field inside a building (top), and representation of the frequency components by spectral analysis (bottom).

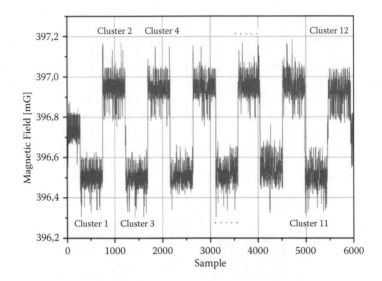

**FIGURE 4.7**
Differential measuring principle used to eliminate interference fields.

at a coil-sensor distance of 12 m. The clusters that result from the switching of the coil current can clearly be seen.

### 4.3.1 Cluster Detection and Digital Filtering

Each cluster contains the captured magnetic field strength during one switching interval. The separation of the signals in clusters is performed by using the cross-correlation with a template signal (square wave signal).

The normalized cross-correlation gives information about the similarity of the two signals and can be computed with the following equation, where $x_1$ and $x_2$ are the two compared signals:

$$R[k] = \frac{\sum_{i=1}^{N} x_1[i] \cdot x_2[i+k]}{\sqrt{\sum_{i=1}^{N} (x_1[i])^2 \cdot \sum_{i=1}^{N} (x_2[i+k])^2}}. \tag{4}$$

In this particular case, the correlation between the captured signal and a template signal is computed and examined in respect to its extreme values (see Figure 4.8). Obviously, a new cluster starts on each of these values (Blankenbach, Norrdine, & Hellmers, 2011).

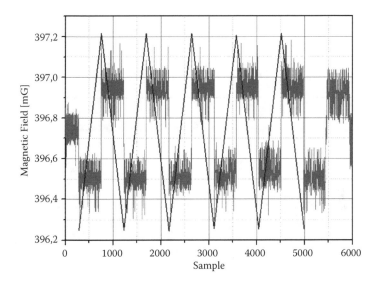

**FIGURE 4.8**
A captured magnetic signal and its cross correlation with a template signal.

The coil's magnetic field $B$ can then be estimated by using the following relation, where $B_t$ is the mean value of a cluster at the time $t$:

$$B = \frac{B_{t+1} - B_t}{2} = \frac{(B + B_{st1}) - (-B + B_{st0})}{2}. \tag{5}$$

$B_{st1}$ and $B_{st0}$ are the overlaying long periodic interference fields. Assuming that $B_{st}$ remains stationary during the switching interval ($B_{st1} = B_{st0}$), the interference would be completely eliminated by calculating the difference as shown in Equation 5. The resulting coil magnetic field can be used for distance determination (cf. Equation 3).

On the basis of the determined single distances from adjacent clusters, the mean distance and the empirically determined variance as a measure of dispersion are calculated. In addition to the long periodic interference fields, high-frequency noise components caused by other electrical devices are superimposed to the measured signal. Since the method of calculating the difference corresponds to a high-pass filter, where its cutoff frequency is related to the switching frequency, the high-frequency noise components in the signal are still present and affect the formation of the cluster mean values. To minimize these noise components, an FIR (finite impulse response) filter with a cutoff frequency of 1 Hz is implemented. Figure 4.9 shows a filtered signal in comparison to the input signal. It can be recognized that the high-frequency noise components are significantly filtered. The output signal is quite similar to the transmitted square wave signal (Blankenbach et al., 2011).

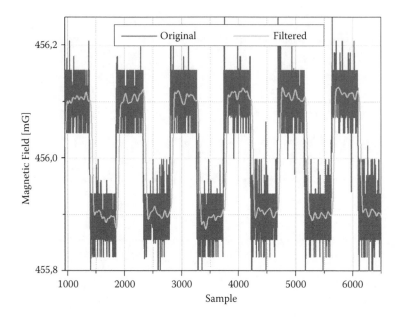

**FIGURE 4.9**
Signal filtering by an FIR low-pass filter.

### 4.3.2 Adaptive Filtering

Although the method of digital filtering provides a helpful method to separate the desired signal from noise, the drawback of regular FIR filters is that they are invariant in respect to time (nonstationary). So it is impossible to be responsive to changing frequency abilities.

To prevent this circumstance at nonstationary signals, we utilized adaptive filters (see Figure 4.10) (Haykin & Widrow, 2003). The adaptive filter was used to find the optimal coefficients ω of a variable filter to extract an estimate of the desired signal, which is in our case an uncorrupted magnetic field at a reference sensor located in a known position.

The adaptive filter is driven by the input signal $x[k]$ (corrupted magnetic field signal at reference sensor) and has as output $y[k]$ (estimate of the desired

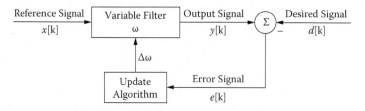

**FIGURE 4.10**
Setup of adaptive filtering.

**TABLE 4.2**

Least Mean Square Algorithm

| | |
|---|---|
| Initialization: | $\omega[0] = \omega_0, \quad \mu > 0$ |
| 1. Filtered signal: | $y[k] = \omega^T[k] \cdot x[k]$ |
| 2. Error value: | $e[k] = d[k] - y[k]$ |
| 3. Iterative process: | $\omega[k+1] = \omega[k] + \mu \cdot e[k] \cdot x[k]$ |

signal). The objective is to change (adapt) the coefficients of the filter in a way that the variance of the error signal $e[k]$ becomes minimal. The computed adaptive filter coefficients will be later used to filter the signal of every sensor in the proximity of the reference sensor. In the case of an indoor positioning system, it is necessary to find the filter coefficients in real time. For this, the least mean square (LMS) algorithm is used. The LMS algorithm was first proposed in 1960 by Widrow and Hoff for use in adaptive switching circuits (Widrow & Hoff, 1960) and is the most popular real-time adaptive filtering technique. It consists in general of two basic processes: the filtering process, which computes the output $y[k]$ of the adaptive filter in response to the input signal $x[k]$ and generates an estimation error $e[k]$ by comparing the output $y[k]$ with the desired signal $d[k]$, and the adaptive process, which adjusts the coefficients vector $\omega[k + 1]$ corresponding to the estimation error $e[k]$ (see Table 4.2) (Blankenbach et al., 2011).

## 4.4 Evaluation of the Experimental System

For the evaluation of the experimental system, practical measurements were conducted in real indoor test environments. In the first step, these measurements were carried out in 2D.

This means the coil and sensor were placed in the same horizontal plane, and thus for performance analysis, the simplified relationship between the sensed magnetic field and distance can be used (cf. Equation 3).

### 4.4.1 Rotational Symmetry

To verify the production quality of the coils with regard to the generated magnetic fields, the coil rotational symmetry was tested by comparing the magnetic field measurements of different rotation angles. Figure 4.11 shows the results of one selected coil. The sensor was located in a horizontal plane at a constant distance of 4 m. After that, measurements were carried out at angular steps of 18 deg around the coil's vertical axis. In the upper part of

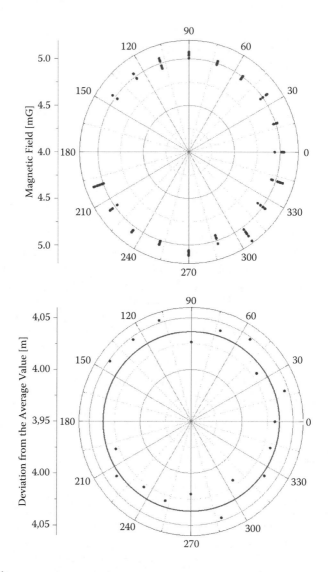

**FIGURE 4.11**
Rotational symmetry of the coil magnetic field.

the figure, the observed magnetic field strengths are plotted for each angular step. The lower part depicts the deviations between the calculated mean distance at each angular step and the overall mean value (circle). In this example, the deviations are between 0.3 cm and 1.7 cm (on average 0.9 cm), which is considered to be sufficient for the rotational symmetry. Since the remaining coils deliver comparable results, rotationally symmetric coil fields are assumed.

### 4.4.2 Maximal Range and Accuracy

The maximal range of the experimental system was tested at direct line of sight (LoS), as well as NLoS, between the coil and the sensor. Figure 4.12 exemplarily shows one measurement scenario. The magnetic coil is located in the middle of the building, and the sensor is placed in different positions along a straight line (Blankenbach, Norrdine, & Hellmers, 2012).

The performance of the current prototype can be shown by plotting the signal-to-noise (SNR) ratio of the magnetic signal relative to the distance between coil and sensor. Figure 4.13 depicts the determined SNR using a coil's current strength of 12 A. In this scenario, the coil and sensor are separated by a 27-cm reinforced concrete wall. At distances between 6 m and 9 m, the sensor is in the vicinity of the home electrical equipment, and furthermore the coil and the sensor are separated by several 24-cm brick walls. As expected, the SNR decreases very rapidly as a function of range and degrades considerably in the proximity of electrical noise sources. However, the presence of obstacles such as walls and/or metallic objects causes essentially no loss in SNR efficiency.

In the present example, the coil field could be detected at distances greater than 15 m. This corresponds to the objective range of 15 m to 20 m and confirms the simulations carried out before.

For analysis of the ranging accuracy, repetitive measurements and comparisons with true distances were performed. As shown in Figure 4.14, in this scenario an accuracy of about 50 cm was achieved in short range (0 m–8 m) and approximately 1 m in the vicinity of the maximum range.

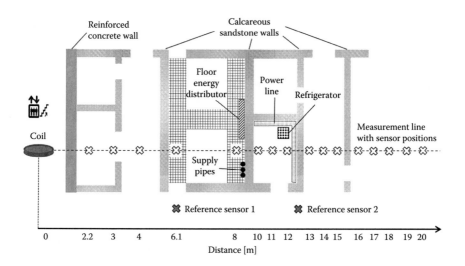

**FIGURE 4.12**
Magnetic coil and sensor locations for a sample NLoS measurement environment.

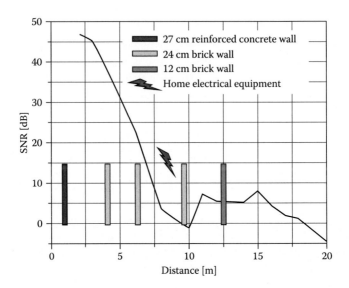

**FIGURE 4.13**
Measured signal-to-noise ratio.

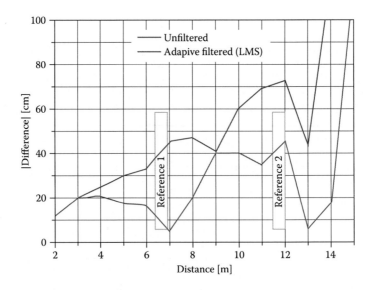

**FIGURE 4.14**
Accuracy comparison between filtered and unfiltered measurements.

By the use of two reference sensors and the adaptive filtering method, the ranging accuracy could be improved to better than 20 cm for short ranges and better than 40 cm for far ranges (8 m–14 m). The standard deviation, as an indicator for the ranging precision, could also be improved to approximately

1 cm and 10 cm in the short range and far range, respectively, by applying the adaptive filter method (cf. Blankenbach et al., 2012).

### 4.4.3 Calibration

The results of the test measurements (see Section 4.2) clearly show an increase of the distance deviation between calculated and true distances with range (= systematic error). Since the positioning principle is based on distance measurements (cf. Section 4.5), calibration measurements of the system (coil-sensor) were conducted in a noise-free environment. These measurements were carried out again first for the 2D case, that is, the coil and sensor were placed in the same horizontal plane. An example of calibration measurements is depicted in Figure 4.15, which shows the deviations between computed and true distances and a best fitting regression line describing the systematic ranging errors. These systematic errors, which were already apparent in the range measurements, can be explained by a model derivation between the measured and the theoretically calculated magnetic field strength (Blankenbach & Norrdine, 2013). Therefore, for each coil a regression line was estimated, whereby the influence of the model deviations could be significantly minimized for the 2D distance determination. The mean value of the resulting residuals is estimated to about 6.5 cm over the total range (1 m–14 m) in the depicted example. For short distances up to 8 m, the residuals are even less than 3 cm on average.

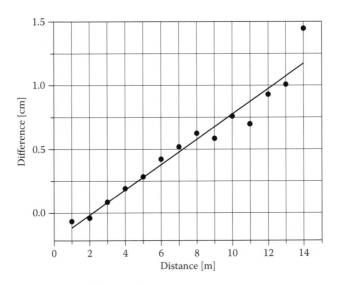

**FIGURE 4.15**
Fitted regression line of distance derivations to one special coil in an electrical noise-free environment.

### 4.4.4 Mass Market (Low-Cost) Sensors

The popularity of smartphones, like the iPhone or Android-based phones, makes them promising platforms for mobile location services inside of buildings, since they contain multiple sensors, including a magnetic field sensor. However, the localization remains a challenge in indoor environments, especially by the use of the embedded low-quality sensors, without additional hardware. An evaluation of the integrated sensors for positioning by using the MILPS was performed. Therefore, the Apple iPhone 3GS, the Apple iPhone 4, and an Android device (Samsung Galaxy Nexus), which contain ultralow power digital three-axis electronic compasses, were tested. Table 4.3 shows the characteristics of the embedded magnetic sensors. Test measurements for determining 2D distances were accomplished by using customized iOS and Android apps (see Figure 4.16) extended by logging and archiving capabilities.

The analysis of the raw data shows that it is possible to detect the coil's magnetic signal up to a distance of 6 m. The ranging accuracy as comparison between the measured and true distances depends on the distance between the coil and the sensor. The deviations to the true distances vary from 0.01 m

**TABLE 4.3**

Magnetic Field Sensors Characteristics

| Smartphone | Device | Manufacturer | Range | Resolution |
|------------|--------|--------------|-------|------------|
| iPhone 3 | AN-203 | Honeywell | ±2 G | 100 µG |
| iPhone 4 | AKM8975 | AKM Semiconductor | ±0,5 G | 100 µG |
| Samsung Galaxy Nexus | YAS529 | Yamaha Corporation | ±8 G | 1.5 mG |

**FIGURE 4.16**
Customized measurement app for Samsung Galaxy Nexus (left) and iPhone 4 (right).

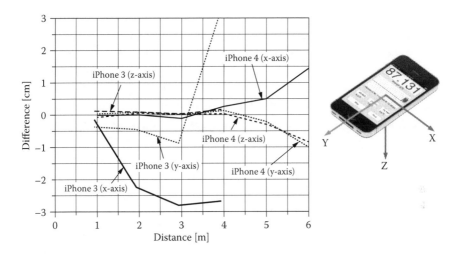

**FIGURE 4.17**
Results of test measurements with iPhone 3GS and iPhone 4.

to 0.02 m for very short distances (1 m–2 m) and from 0.1 m to 0.15 m for distances between 2 m and 6 m. For iPhone 3GS, it has to be noted that the maximal range is about 5 m, and the mentioned accuracy is reachable only if the z-axis of the sensor points toward the coil's center (see Figure 4.17). This could be caused by different axis sensitivities of the built-in sensor. The comparison between iPhone 4 and iPhone 3GS shows that iPhone 4 provides better performances than iPhone 3GS, on account of a more sensitive magnetic sensor. This could also be verified by the estimation of the measuring rate and resolution of the raw data of both phones (Blankenbach et al., 2011).

## 4.5 Position Estimation

The position estimation with the MILPS relies on the trilaterion principle that is described in the following section. Therefore, a distinction is made between 2D and 3D positioning.

### 4.5.1 2D Position Estimation

*Algorithm:* As mentioned earlier, the measured magnetic field *B* depends only on the distance *r* between the sensor and the coil center, if the coil and the magnetic field sensor are approximately located in the same horizontal plane (see Equation 3). Thus, three horizontal distances can be derived from the observed field strengths of three coils. Using the known coordinates of the reference coils $(x_i, y_i)$, the sensor position is calculated by means of trilateration. In geometrical terms two possible positions already result from the intersection of two circles. Each circle center is represented by a coil, and the circle radiuses are defined by the corresponding horizontal distances to the sensor. By comparing the observed distance to the third coil with the calculated distances of the two potential positions, one position can be rejected.

The calculated solution $(x_0, y_0)$ is then used as an initial guess for a Gauss-Newton estimation, which solves the linearized equations of

$$r_{0i} = \sqrt{(x_0 - x_i)^2 + (y_0 - y_i)^2} \tag{6}$$

by least squares adjustment.

Considering all distances lead to the following equation system for the position estimation

$$r + v = A \cdot \Delta x \tag{7}$$

where $r$ describes the vector of distance measurements $r_i$ from sensor to coil $i$, $v$ is the vector of residuals $v_i$, $A$ is the Jacobian matrix as a result of linearization of the nonlinear observation Equation 6 at the starting position, and $\Delta x$ describes the vector of unknowns as parameter additions to the starting point (see Equation 8) (Blankenbach et al., 2012).

$$\begin{pmatrix} r_1 \\ r_2 \\ \vdots \\ r_n \end{pmatrix} + \begin{pmatrix} v_1 \\ v_2 \\ \vdots \\ v_n \end{pmatrix} = \begin{pmatrix} \dfrac{\partial r_1}{\partial x_p} & \dfrac{\partial r_1}{\partial y_p} \\[2mm] \dfrac{\partial r_2}{\partial x_p} & \dfrac{\partial r_2}{\partial y_p} \\[2mm] \vdots & \vdots \\[2mm] \dfrac{\partial r_n}{\partial x_p} & \dfrac{\partial r_n}{\partial y_p} \end{pmatrix} \cdot \begin{pmatrix} \Delta x_p \\ \Delta y_p \end{pmatrix} \tag{8}$$

Considering that the observed distances are subject to uncertainties expressed by a variance $\sigma_i^2$, a covariance matrix $Q_{ll}$ can be used for deriving observation weights $p_i$:

$$Q_{ll} = diag(\sigma_1^2, \ldots, \sigma_n^2) \tag{9}$$

$$P = Q_{ll}^{-1} = diag(p_1, \ldots, p_n) \tag{10}$$

The least squares solution finally results from

$$x = (A^T P A)^{-1} A^T P l \qquad (11)$$

By using the estimates in $x$ for calculating the starting point for the next iteration, the Gauss-Newton algorithm finally delivers the desired coordinates and a quality measure from the covariance matrix of unknowns (cf. Niemeier, 2001).

*Measurement experiments with an industrial sensor:* 2D position determination was carried out by using the magnetic field measurements of three different coils with an applied current of 12 A at the university building (see Figure 4.18). The reference stations (coils) and the MS (HMR2300) were placed in the same horizontal plane. The test field comprises several rooms and corridors separated by walls and doors. Beside the MS, which was again placed along a straight line, a reference sensor was placed at a known position in order to use the method of adaptive filtering (cf. Section 4.3.2). Figure 4.18 shows the location of the coils, as well as the MS and the reference sensor, in different rooms. For distance determinations, the adaptively filtered signals were used with and without calibration (see Section 4.4.3) to reduce the systematic errors. Figure 4.18 shows the positions calculated from uncorrected distances and from corrected distances in comparison to the true positions.

For the corrected distances, the empirical standard deviations determined from the average values after cluster detection (see Section 4.3.1) are used for building the covariance matrix of observations. Because there is no reliable covariance information for the uncorrected distances, it is assumed that the observations in this case are equally corrupted by noise, and therefore all observations get the weight $p_i = 1$. The evaluation of the localization results shows that calibrated observations deliver positions that are closer to true

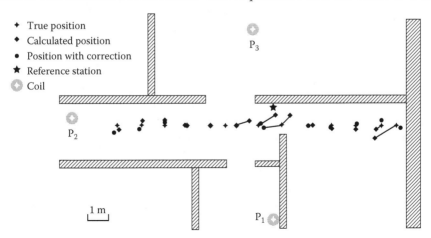

**FIGURE 4.18**
2D positioning results.

positions than the noncalibrated observations. While the position deviations in the first case stand in the range of 30 cm, the position deviation in the second case with calibration is on average less than 12 cm.

*Positioning with smartphones:* In addition to the high-quality industrial sensor HMR2300, the ability to determine a local position in an indoor environment was accomplished with iPhone 4. As previously stated, the coil's magnetic field can be detected only up to a range of 6 m because of the lower resolution of the embedded sensors (see Section 4.4.4). Hence, the positioning experiments were performed in a 7 m × 7 m test field. Figure 4.19 shows the position plan of a localization experiment with iPhone 4 as MS.

The MS and all coils were placed in the same horizontal plane to obtain a two-dimensional position using the algorithm explained before. Because at present time there exists no calibration information for iPhone 4, all distance observations are equally weighted ($p_i = 1$) within the least squares estimation.

Table 4.4 shows the iPhone 4 positioning results. Pointed out are the deviations to the true values in respect to x- and y-coordinates, as well as position deviations. At three locations (1, 5, and 8) it was not possible to get a unique solution, since not all three coils could be reliably detected. In all other cases, the accuracy of the coordinates is better (or equal to) than 0.5 m (excluding

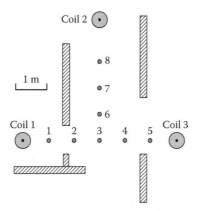

**FIGURE 4.19**
Magnetic coils and sensor locations for iPhone 4 2D positioning test.

**TABLE 4.4**

Results of iPhone 4 2D Positioning Measurements

| Position | 1 | 2 | 3 | 4 | 5 | 6 | 7 | 8 |
|---|---|---|---|---|---|---|---|---|
| dX [cm] | — | 52 | 12 | −24 | — | −9 | −9 | — |
| dY [cm] | — | 3 | 76 | 4 | — | −27 | −19 | — |
| dP [cm] | — | 52,1 | 76,9 | 24,3 | — | 28,5 | 21,0 | — |

the y-coordinate of position 3), which demonstrates a satisfactory result for indoor positioning with a low-cost sensor.

### 4.5.2 3D Position Estimation

Besides the 2D positioning, a 3D position estimation algorithm was developed. One approach, already introduced in Prigge (2004) and adapted for the MILPS, is based on the solution of the three extended magnitude equations:

$$B_i = k \frac{\sqrt{3(z-z_i)^2 + r_i^2}}{r_i^4} \text{ with } k = \mu_0 \frac{NIF}{4\pi} \tag{12}$$

Similar to the equations in Section 4.5.1, the resulting equation system (Equation 12) can be solved by using the Gauss-Newton method, whereby for finding a starting point, some assumptions have to be made. A more detailed description can be found in Prigge (2004) and Blankenbach et al. (2012).

Because the experimental system consists only of three coils, the minimum number of three distance observations has to be used for 3D positioning. Because of this lack of redundant observations, another approach was developed for 3D position estimation with the MILPS (see Figure 4.20). In the initialization step, it is assumed that the elevation angles $\theta_i$ are equal to zero for all field strength observations, so that Equation 3 can be utilized for distance derivation. Because this assumption always delivers distances that are longer than (or equal to) the true horizontal distances, a sphere intersection can be applied for a 3D position estimate by using a direct solution algorithm (cf. Blankenbach & Willert, 2009). In the next step, this first guess is recursively utilized for the calculation of $\theta_i$, which in turn can be substituted for an improved calculation of the slope distances $r_i$. These new distances are then used for the next position estimation. The iteration repeatedly recalculates a new position estimate until no significant change of coordinates is recognizable, which leads to the final estimate $(x_{MS}, y_{MS}, z_{MS})$ for the 3D position. Because the direct solution always delivers two potential solutions, one position has to be rejected because of an assumption, for example, regarding

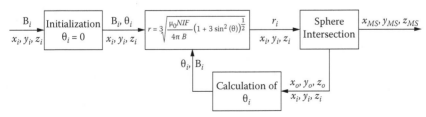

**FIGURE 4.20**
Iterative algorithm for 3D position estimation with the MILPS.

the z-component. Furthermore, this approach does not consider observation weights since the system 3D calibration is still to be made.

*Measurement experiments:* To evaluate the 3D position estimation, we examined the positioning accuracy by comparing the true and calculated position in different measurement experiments.

An example of 3D position determination was carried out by using magnetic field measurements of three different coils ($I = 12$ A) at the university building. Figure 4.21 depicts the location of the coils and the MS in different rooms. The three coils and the sensor were placed at different heights above ground ($h_{coil1} = 1.14$ m, $h_{coil2} = 1.79$ m, $h_{coil3} = 4.23$ m, and $h_{sensor} = 1.12$ m). Table 4.5 shows the results as deviations between the calculated coordinates and the true coordinates based on the raw distances, which means without adjusting the derived distances (e.g., by using calibration functions).

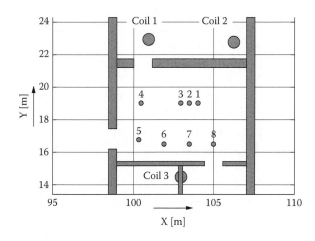

**FIGURE 4.21**
Magnetic coils and sensor locations for 3D positioning test.

**TABLE 4.5**

3D Positioning Results

| Position | dX [cm] | dY [cm] | dZ [cm] |
|:---:|:---:|:---:|:---:|
| 1 | −26 | 11 | 7 |
| 2 | −26 | 8 | 14 |
| 3 | −22 | 8 | 13 |
| 4 | −7 | 1 | 13 |
| 5 | −6 | 22 | 12 |
| 6 | −29 | 19 | 24 |
| 7 | −46 | 15 | −6 |
| 8 | −42 | 5 | −43 |

In another experiment, the approach was evaluated in a small-scale scenario (4 m × 4 m × 3 m) to enable signal detection by smartphones with an applied coil current of 12 A (see Figure 4.22). Table 4.6 shows the results as deviations between the determined coordinates and the prior known coordinates. Thereby it has to be noted that the coils and sensors are located almost in the same horizontal plane. It was not applied calibration. If we place the sensor on the same horizontal plane, then it is not a 3D positioning more. Thus, because of the suboptimal coil's vertical configuration, the intersection conditions are poor for the vertical component. However, as expected the HMR2300 provides the best performance with accuracy in the centimeter range (disregarding the z-component).

Using low-cost magnetic sensors embedded in smartphones, accuracies in the dm-range can be expected. In this scenario, the iPhone 4 device provides better positioning accuracy than the Samsung Galaxy Nexus, probably because of a more sensitive magnetic sensor.

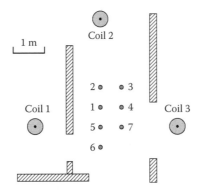

**FIGURE 4.22**
Location plan of magnetic coils and sensor for 3D positioning test.

**TABLE 4.6**

3D Positioning Results

| | HMR2300 | | | iPhone 4 | | | Samsung Galaxy Nexus | | |
|---|---|---|---|---|---|---|---|---|---|
| Position | dX [cm] | dY [cm] | dZ [cm] | dX [cm] | dY [cm] | dZ [cm] | dX [cm] | dY [cm] | dZ [cm] |
| 1 | −3 | 0 | −19 | −3 | −2 | 46 | 9 | −9 | −141 |
| 2 | −3 | 0 | 11 | 2 | −12 | −82 | 7 | −26 | −130 |
| 3 | −4 | −3 | −15 | −6 | −4 | −45 | −2 | −20 | −119 |
| 4 | −4 | −1 | 14 | −6 | 0 | 45 | −13 | 0 | 96 |
| 5 | −3 | −0 | 10 | −3 | 3 | 62 | 10 | −11 | −143 |
| 6 | −1 | −6 | −100 | 5 | −14 | −131 | 12 | 11 | −142 |
| 7 | 10 | −8 | −57 | −10 | −3 | 52 | −12 | 8 | 82 |

## 4.6 Conclusion and Outlook

The results of the measurements performed with the prototype confirm the feasibility to determine the three-dimensional position of a user or object in the building even in NLoS conditions based on artificially generated magnetic fields.

Using the experimental system (three coils and an industrial triaxial magnetic sensor), 2D positioning accuracies of better than 0.5 m can be achieved in laboratory scale (15 m × 15 m). Considering a simple calibration approach, the accuracy of 2D localization could be improved significantly to better than 0.1 m. By means of digital signal processing, the influence of interference fields can be minimized. Consequently, the accuracy of position estimation can be improved. At short ranges even smartphones can be utilized for user localization in an indoor environment.

3D positioning accuracies in the range of 0.5 m in a 10 m × 10 m × 6 m field could be achieved. However, the accuracy could be improved by 3D calibration in future work. In a smaller 4 m × 4 m × 3 m measuring volume, 3D positioning accuracies better than 0.2 m in x- and y-coordinates (and 1.5 m in z-coordinate) could be achieved by using iPhone 4 and Samsung Galaxy Nexus.

Based on the experimental system, a 1 m diameter coil was fabricated, enabling the use of higher currents. Measurements at higher currents lead to a significant increase in range and thus enable building-wide coverage with a small number of coils. On the basis of this application, scenarios are conceivable in which the coil's infrastructure is deployed on demand, for example, in front of a building for the localization of safety or rescue teams.

Future work includes, besides the improvement of 3D position estimation (particularly the 3D calibration), further development with regard to kinematic applications. The latter has been already implemented by sensor fusion (cf. Hellmers, Norrdine, Blankenbach, & Eichhorn, 2013) and by using embedded smartphone sensors (cf. Real Ehrlich & Blankenbach, 2014).

### Acknowledgments

The authors would like to gratefully thank the German Research Foundation (DFG) for supporting the research project (Grant No. BL 1092/1-1) (http://www.gia.rwth-aachen.de/IndoorPositionierung_MILPS/).

# References

Anderson, P. T. (1995). *Pulsed-DC position and orientation measurement system*. U.S. Patent No. 5453686.

Blankenbach, J., & Norrdine, A. (2013). Indoor-Positionierung mit künstlichen Magnetfeldern Von der Innenraumpositionierung zu standortbezogenen Diensten in Gebäuden, ZfV. Heft 01/13.

Blankenbach, J., Norrdine, A., & Hellmers, H. (2011). Adaptive signal processing for a magnetic indoor positioning system. In *2011 international conference on indoor positioning and indoor navigation (IPIN)*. New York, NY: IEEE.

Blankenbach, J., Norrdine, A., & Hellmers, H. (2012). A robust and precise 3D indoor positioning system for harsh environments. In *2012 international conference on indoor positioning and indoor navigation (IPIN)*. New York, NY: IEEE.

Blankenbach, J., & Willert, V. (2009). Robuster Räumlicher Bogenschnitt: Ein Ansatz zur robusten Positionsberechnung in Indoor-Szenarien. AVN, Heft 8/9.

Blood, E. B. (1990). *Device for quantitatively measuring the relative position and orientation of two bodies in the presence of metals utilizing direct current magnetic fields*. Patent No. 4849692.

Callmer, J., Skoglund, M., & Gustafsson, F. (2010). Silent localization of underwater sensors using magnetometers. *EURASIP Journal on Advances in Signal Processing*, *2010*, Article ID 709318.

Dong, L., Hanwen, L., Wentao, S., Youyun, X., Xingzhao, L., & Wenjun, Z. (2004). 3D positioning system using magnetic field. In *Fifth world congress on intelligent control and automation, 2004. WCICA 2004* (Vol. 4, pp. 3668–3670). New York, NY: IEEE.

Haykin, S., & Widrow, B. (Eds.). (2003). *Least-mean-square adaptive filters*. Hoboken, NJ: John Wiley & Sons.

Hellmers, H., Norrdine, A., Blankenbach, J., & Eichhorn, A. (2013). An IMU/magnetometer-based indoor-positioning system using Kalman filtering. In *2013 international conference on indoor positioning and indoor navigation (IPIN)*. New York, NY: IEEE.

Hu, C., Meng, M.Q.-H., Mandal, M., & Wang, X. (2006). 3-axis magnetic sensor array system for tracking magnet's position and orientation. In *The sixth world congress on intelligent control and automation, 2006* (Vol. 2, pp. 5304–5308). New York, NY: IEEE.

Kuipers, G. (1975). *Object tracking and orientation determination means, system and process*. U.S. Patent No. 3868565.

Lehner, G. (2006). *Elektromagnetische Feldtheorie für Ingenieure und Physiker* (5th ed.). Berlin, Heidelberg, New York: Springer-Verlag.

Li, B., Gallagher, T., Dempster, A. G., & Rizos, C. (2012). How feasible is the use of magnetic field alone for indoor positioning? In *International conference on indoor positioning and indoor navigation (IPIN)* . New York, NY: IEEE.

Li, M., Song, S., Hu, C., Yang, W., Wang, L., & Meng, M.Q.-H. (2009). A new calibration method for magnetic sensor array for tracking capsule endoscope. In *IEEE international conference on robotics and biomimetics (ROBIO)* (pp. 1561–1566). New York, NY: IEEE.

Niemeier, W. (2001). *Ausgleichungsrechnung (de Gruyter Lehrbuch)*. Berlin, Germany: De Gruyter.

Paperno, E., Sasada, I., & Leonovich, E. (2001). A new method for magnetic position and orientation tracking. *IEEE Transactions on Magnetics*, 37(4), 1938–1940.

Prigge, E. A. (2004). *A positioning system with no line-of-sight restrictions for cluttered environments* (Unpublished thesis). Stanford University, Stanford, CA.

Prigge, E. A., & How, J. P. (2004). Signal architecture for a distributed magnetic local positioning system. *IEEE Sensors Journal*, 4(6), 864–873.

Raab, F. H. (1982). *Remote object position and locater*. U.S. Patent No. 4314251.

Raab, F. H., Blood, E. B., Steiner, T. O., & Jones, H. R. (1979). Magnetic position and orientation tracking system. *IEEE Transactions on Aerospace and Electronic Systems*, 15(5), 709–718.

Real Ehrlich, C., & Blankenbach, J. (2014). *Innenraumpositionierung für Fußgänger unter Verwendung eines Smartphones. Geoinformationssysteme*. Berlin, Germany: Wichmann Verlag.

Subbu, K. P., Gozick, B., & Dantu, R. (2013). LocateMe: Magnetic-fields-based indoor localization using smartphones. *ACM Transactions on Intelligent Systems and Technology (TIST)*, 4(4), Article 73.

Widrow, B., & Hoff, M. E. (1960). Adaptive switching circuits. In *IRE WESCON convention record* (Part 4, pp. 96–104). Los Angeles, CA: Western Electronic Show and Convention.

# 5

## Localization in Underground Tunnels

Fernando Pereira

Adriano Moreira

Christian Theis

Manuel Ricardo

### CONTENTS

*Abstract:* Positioning technologies are becoming ubiquitous and are being used more and more frequently for supporting a large variety of applications. For outdoor applications, global navigation satellite systems (GNSSs), such as the global positioning system (GPS), are the most common and popular choice because of their wide coverage. GPS is also augmented with network-based systems that exploit existing wireless and mobile networks for providing positioning functions where GPS is not available or to save energy in battery-powered devices. Indoors, GNSSs are not a viable solution, but many applications require very accurate, fast, and flexible positioning, tracking, and navigation functions. These and other requirements have stimulated research activities, in both industry and academia, where a variety of fundamental principles, techniques, and sensors are being integrated to provide positioning functions to many applications.

The large majority of positioning technologies is for indoor environments, and most of the existing commercial products have been developed for use in office buildings, airports, shopping malls, factory plants, and similar spaces. There are, however, other spaces where positioning, tracking, and navigation systems play a central role in safety and in rescue operations, as well as in supporting specific activities or for scientific research activities in other fields. Among those spaces are underground tunnels, mines, and even underwater wells and caves.

This chapter describes the research efforts over the past few years that have been put into the development of positioning systems for underground tunnels, with particular emphasis in the case of the Large Hadron Collider (LHC) at CERN (the European Organization for Nuclear Research), where localization aims at enabling more automatic and unmanned radiation surveys.

Examples of positioning and localization systems that have been developed in the past few years for underground facilities are presented in the following section, together with a brief characterization of those spaces' special conditions and the requirements of some of the most common applications. Section 5.2 provides a short overview of some of the most representative research efforts that are currently being carried out by many research teams around the world. In addition, some of the fundamental principles and techniques are identified, such as the use of leaky coaxial cables, as used at the LHC. In Section 5.3, we introduce the specific environment of the LHC and define the positioning requirements for the envisaged application. This is followed by a detailed description of our approach and the results that have been achieved so far. Some last comments and remarks are presented in a final section.

## 5.1 Positioning in Underground Tunnels and Similar Environments

While it is true that research and development of positioning systems are now a few decades old, it was in the past decade that we observed a significant increase in the intensity of research activities and in the availability of commercial products in this area. Besides GPS and GLONASS, and the efforts around the announced Galileu GNSS system, which provide global positioning systems mostly outdoors, we have observed the widespread use of wireless and cellular network-based positioning systems through their integration into smartphones, tablet computers, and in-car navigation systems. For indoor environments, pioneer work developed at the Olivetti Research Laboratory in Cambridge, England, by Andy Hopper and his team, the Active Badge System, started a new era in the development of positioning technologies (Want, Hopper, Falcão, & Gibbons, 1992). The RADAR system, developed by P. Bahl and V. N. Padmanabhan at the Microsoft Research Labs, exploited the characteristics of the then emerging wireless local area networks (WLANs) to demonstrate the feasibility and usefulness of positioning systems for indoor spaces (Bahl & Padmanabhan, 2000). The fundamental concepts introduced by these works are still being used in many of the most recent approaches for indoor positioning.

For indoor applications, most of the existing and proposed positioning solutions are targeted at well-controlled environments used by humans in their daily life such as office buildings, shopping malls, train stations, hospitals, and warehouses. For these contexts, there is a large variety of positioning solutions, with different requirements, and they are based on many different techniques and sensors: pseudolites, radio fingerprinting, ultrasound, laser ranging, inertial navigation, or even pressure sensors, just to name a few. However, much less research has been conducted for developing positioning, tracking, and navigation systems for underground spaces.

Underground spaces include tunnels of various kinds, mines, and even water-filled wells and tunnels. Examples of these environments include metropolitan train tunnels, car tunnels, water and waste drain tunnels, mines, and natural caverns. Most of the applications for these environments are related to the safety of workers or to the tracking and navigation of rescue personnel in catastrophes. Therefore, these applications often demand special functionality and pose very demanding requirements for their development, deployment, and operation.

Tunnels and other underground facilities are often environments of extreme conditions, dangerous for people, and very demanding for the deployed positioning systems. Mines are probably one of the most challenging kinds of underground environments because of their dynamic characteristics. Among the characteristics of tunnels, the following have a great impact on the performance of positioning systems (Mautz, 2012; Misra, 2009):

- High attenuation for GNSS (GPS) signals, making GNSS unusable in the large majority of cases, including most car tunnels;
- Difficult conditions for the propagation of radio signals inside the tunnels such as high attenuation, reflections, refraction, and multipath fading;
- High temperatures, especially in mines and other spaces where large machinery is used;

- High humidity that affects the propagation of radio and ultrasound signals;
- The presence of hazardous gases, often leading to the high risk of explosions;
- Difficulty accessing electric energy sources to power the positioning infrastructure; and
- The difficulty, or even the impossibility, of installing new equipment for supporting the positioning system.

Given these conditions, each envisioned application leads to a particular set of requirements. For the particular case of underground constructions, Oliver Schneider (2010) identified a set of requirements for positioning and navigation, including range, accuracy, 3D positioning capabilities, toughness, user friendliness, power availability, and cost, among others. Other particular systems include these and additional requirements, most of which are not usually considered for general-purpose positioning systems. For rescue operations, being infrastructure-free is often among the most important requirements.

Besides the construction of tunnels (Schneider, 2010), other underground contexts have been considered for the development of positioning systems. Fink, Beikirch, Voss, and Schroder (2010) proposed a positioning system, based on RSSI (received signal strength indicator) fingerprinting and inertial navigation, to improve the placement of machinery in longwall coal extraction mines. Fairfield, Kantor, and Wettergreen (2007) reported a simultaneous localization and mapping (SLAM) system for scientific exploration in water-filled tunnels and wells. It uses a particle filter to combine input data from several types of sensors, allowing an autonomous underwater vehicle to create a map of its surroundings and use it to navigate through the environment. Julia Engelbrecht (Engelbrecht, Forster, Michler, & Collmann, 2012) presented another interesting solution for the positioning of mobile devices inside a tram vehicle (a kind of a tunnel, although not underground). Also not specifically underground but under extreme conditions is the system described by Phithakkitnukoon et al. (2013) for tracking trash along its removal chain.

The techniques used for these particular environments and for these specific applications are of many types. Some of them are briefly introduced in the next section.

## 5.2 Sensors and Techniques for Positioning and Tracking

Many of the techniques that have been proposed for positioning inside office buildings and similar spaces are also being explored for positioning in underground tunnels. Most of them rely on the use of one or more sensors to measure some characteristics of the surrounding environment or how signals propagate in space. These signals are then processed using range-based time of flight, angle of arrival, or scene analysis approaches, together with geometric algorithms, to produce an estimate for the position of a given device. For a complete survey of indoor positioning technologies, see Mautz (2012). Most of the existing positioning systems for licensed indoor use are 2D, meaning that the position of the device or person is estimated on a two-dimensional Cartesian coordinate system. These systems are often evaluated based on their accuracy, defined as the average positioning error or as the maximum expected error on a given percentage of the estimates. The error is, in this case, the Euclidean distance between the estimated positions and the real positions. For multifloor buildings, the position estimates are usually described by a pair of coordinates (x,y) plus a floor identifier, leading to what are known as 2.5D systems. Some systems are able to provide real 3D position estimates, and these are often very accurate. At the other extreme are systems that provide position estimates along a line, such as the distance from the entry point on a tunnel. These are 1D positioning systems. Most of the solutions for tunnels are 1D, but 3D systems also exist for underground positioning. Some examples of these systems are described next.

The Infrasurvey is a 3D positioning system based on the magnetic fields generated by mobile underground transmitters and detected by receivers at the surface.[*] Orientation of the transmitters can also be estimated. According to Infrasurvey, the position and orientation of the transmitters can be estimated at up to 300 meters below the surface with an accuracy of 1 meter by taking advantage of the low absorption of magnetic fields by rocks.

Mines are one of the underground contexts for which many positioning systems have been proposed. Most of these systems are based on zone or proximity detection (Misra, 2009). Aiming to develop a positioning system for the extreme conditions of underground mines, Prasant Misra proposed a solution based on a wireless sensor network and ultrasound ranging in order to provide positioning services without the need of a fixed infrastructure (Misra, 2009). Another solution for positioning inside coal extraction longwall mines was proposed by Andreas Fink (Fink et al., 2010), aiming to automate the placement of machinery. This solution combines RSSI-based positioning and inertial navigation to increase the positioning accuracy to better than 1 meter.

Safety in narrow-vein mines can also be improved by deploying positioning systems for localizing and tracking workers and equipment. For these scenarios, Dayekh, Affes, Kandil, and Nerguizian (2011) proposed a system based on cooperative spatio-temporal diversity where fingerprints are collected from multiple access points at different time instants and combined using artificial neural networks (ANNs) to estimate the 1D position of

---

[*] See Infrasurvey, http://www.infrasurvey.ch/.

moving objects. This cooperative approach is claimed to outperform previous solutions, also based on ANNs, that use spatial or temporal diversity only. The reported results point to an accuracy of better than 25 cm for 90% of the estimates.

Some solutions have also been proposed for car and train tunnels. The use of pseudolites has been proposed for tunnels where GPS signals are unavailable, to provide positioning at centimeter-level accuracies (Michalson, 2000). In this work, the tunnel geometry is taken into consideration to eliminate the undesired effects of multipath on the propagation of the radio signals. Centimeter-level accuracies can be achieved if appropriate places for positioning the antennas in straight tunnels are selected. Other tunnel geometries, such as curved tunnels, require further study.

The use of RFID (radio frequency identification) technology combined with GPS and gyroscopes has also been exploited to improve the positioning and navigation of vehicles in road tunnels and downtown areas (Chon, 2005). The basic idea relies on embedding RFID tags into the pavement of roads and on the installation of RFID readers in the vehicles. The information obtained by the readers when a tag is detected on a vehicle is then combined with GPS and gyroscope data to produce very accurate positioning estimates.

Because of the geometry of tunnels, leaky coaxial cables—or *leaky-feeders*—are particularly adequate to provide uniform radio communication services. Recently, these cables have also been considered for supporting positioning systems. Nakamura et al. (2010) proposed a positioning system based on a leaky coaxial cable for tracking rescue workers in tunnels and passages in underground facilities. The basic idea is to inject a nonmodulated pulse at a carrier frequency of 2.4 GHz into the cable, let the tracked devices receive these pulses, and down-convert them to 1.2 GHz pulses that are transmitted back to the cable. The position along the cable is estimated from the time difference between the transmitted pulses and the received pulses. An accuracy of about 1 meter over a section 100 meters long has been reported, but larger errors, up to 10 meters, have been observed in some sections of the cable.

Other researchers have also reported the use of leaky coaxial cables for indoor positioning (Engelbrecht, Collmann, Birkel, & Weber, 2011; Engelbrecht et al., 2012; Weber, Birkel, Collmann, & Engelbrecht, 2010; Weber, 2011). Weber et al. (2010) evaluated the performance of radio frequency fingerprinting over leaky coaxial cables using several alternative positioning algorithms. The reported results show that an accuracy of around 8 meters (at 80% probability) can be obtained with standard leaky coaxial cables. A new type of cable, particularly adapted for positioning, has also been proposed. Engelbrecht et al. (2012) proposed leaky coaxial cables to support positioning inside public transportation trams. Experimental results obtained along a 9-meters-long wagon of a tram, with a leaky coaxial cable deployed on the wagon ceiling, prove the feasibility of this approach for positioning, with accuracies around 1.3 meters (95.45% confidence).

## 5.3 Localization and the LHC Tunnel

CERN is the world's largest particle research laboratory, and it sits in the Franco-Swiss border close to Geneva. To accelerate particles, very complex machines, called accelerators, have been developed since the early days of CERN, which have been continuously updated and extended. In the context of CERN activities, localization in the vast underground areas has become of great importance, mostly for safety reasons. In such a controlled and hazardous environment, it is extremely helpful to know the position of workers to provide faster assistance in case of an emergency. From users' perspective, localization technologies would also play a central role for workers' orientation in the tunnel, so that they could be sure about their own position and take the best way to a certain region while avoiding blocked or dangerous paths. Although the localization accuracy for safety purposes does not necessarily have to be very high, other applications could greatly benefit from higher positioning resolutions. For instance, location-aware systems could identify their current location and label sensed data accordingly or even to change operational modes. This would enable myriad applications for the diverse departments at CERN.

### 5.3.1 The LHC Characteristics

The LHC machine is the last stage in the accelerator chain. It is a very large ring accelerator installed 100 m below the surface and measures 27 km in circumference and accelerates protons to 99.9999991% of the speed of the light in two opposite directions. At four specific points in the accelerator, particles are made to collide, and with the help of massive particle detectors, physicists record and analyze the collisions, searching for phenomena that occur only when such high levels of energy are available, including the creation of unknown elementary particles. Among the most notable achievements, the LHC permitted the observation of the Higgs boson, which consolidates the theory of the standard model and granted François Englert and Peter W. Higgs the 2013 Nobel Prize in Physics.

Although they seem to obey a perfectly circular shape, the LHC tunnel sections can be either completely straight or slightly bent. The LHC tunnel is, in fact, an arched tunnel that follows a standard layout in most of its extension. Except at very specific points, like the experiment caverns, a typical tunnel cross-section has a 2.2 m radius, according to the layout as shown in the right side of Figure 5.1. In the LHC tunnel, besides the LHC machine itself, a large amount of auxiliary equipment exists, including a massive cryogenics cooling system, electronics, and trays of cabling. Although this tunnel layout might look relatively ideal, tunnels and particularly the existing scenario present many adverse physical characteristics and limitations for radio signals, such as the following:

- *Extreme path-loss:* High losses are due to material absorption and humidity.
- *Reflections and refractions:* Because of the irregular nature of the walls and the presence of obstacles, distorted reflections turn into noise.

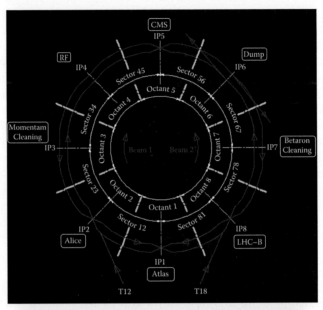

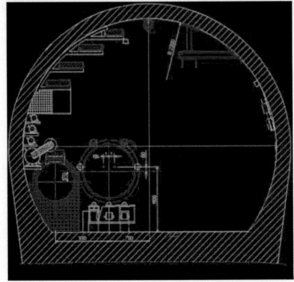

**FIGURE 5.1**
The layout of the Large Hadron Collider tunnel (top), and a schematic view of a tunnel cross-section (bottom).

- *Multipath fading:* The random combination of multiple propagation paths leads to fast variations in the signal strength. Hence, this effect is known as *fast-fading*.
- *Variable propagation velocity:* Tunnels act as waveguides, whose speed of propagation is lower relatively to air and subject to variations due to temperature changes.
- *Noise:* Electromagnetic noise in the environment will interfere with communications, in particular noise caused by motors, power lines, and appliances since they operate in nearby frequency bands.

Specific to the LHC, to avoid interference with the existing control systems, the installation of hardware and emission of radio frequency waves is subject to very restrictive control procedures. Network coverage is available all along the tunnel's length via a set of leaky coaxial cables, which are installed at a distance of nearly 2 m from the ground. Several GSM channels are injected in the cable at different points, and in general, most parts of the tunnel are covered with two GSM channels while several traces of other channels can still be detected. In this configuration, if someone is walking along the tunnel, one of the radio channels gets stronger while the other attenuates. Per the specification, at 900 MHz the cable introduces a longitudinal loss of 3.16 dB/100 m. Other networks are available in the LHC, some of which are expected to be deployed in the tunnel at a later time. First, WLAN will be available in some predefined areas for bit-rate demanding applications during maintenance periods. Second, Terrestrial Trunked Radio (TETRA) will be deployed over the same leaky-feeder serving GSM thoroughly and permanently in the tunnels as part of the safety plans.

### 5.3.2 Applications of Localization for Radiation Protection

Besides the application of locating people as part of a safety plan, localization services would have the potential to help build more automatized systems and procedures. In a scenario involving radiation, a faster and more unmanned process translates into less exposure to the hazard and outstanding advantages in terms of risk prevention and minimization. Of particular interest to the radiation protection group are the frequent radiation surveys conducted throughout the entire accelerator complex. They require radiation measurements in thousands of points whose position must be accurately identified in order to build the radiological profile of the existing equipment.

These surveys are of extreme importance for the safety of personnel, but, nevertheless, this task is performed in a rather simplistic and manual fashion. In the basic old-fashioned way, the radiation survey teams, equipped with probes, measure the radiation levels at specific points and write these values down into forms. This simple task requires many hours by workers in determining the location and filling in the forms in the underground area. In addition, given the amount of data, processing is rather limited, and statistical analysis becomes therefore impracticable. Within this group, the aforementioned problem had been subject of analysis for wide optimization, giving birth to the Radiation Logging project. The goal of this project is to design and implement a logging system for radiation surveys, applying state-of-the-art techniques for data acquisition, transmission, storage, and retrieval.

In Figure 5.2, an overview of the Radiation Logging project is illustrated. The radiation data measured by the detectors is expected to be collected to a mobile computer or tablet, which ensures the validity of the data and automatically determines the position of the measurement. This device provides an interface so that a person might supervise the process. The information is then transferred to a central database while eventually replicated for both safety and availability reasons. In the end, a front-end interface provides the user with the tools to browse and create data analysis reports.

The Radiation Logging project introduces a wide range of matters and challenges from very diverse research areas, which are beyond the scope of this chapter. The following sections present the research in localization that was carried out specifically for the LHC.

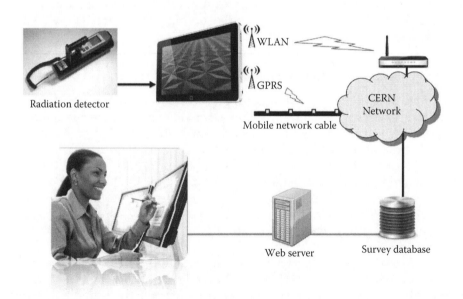

**FIGURE 5.2**
An overview of the Radiation Logging project.

## 5.4 The Technologies Developed for the LHC Tunnel

Many parameters were analyzed technologies were selected for localization in such a special scenario. Besides the technical difficulties the medium presents by itself and the risks of interference as mentioned before, the high levels of radiation present in the LHC turn any solution requiring installation of local hardware virtually senseless. Nonspecialized hardware would burn after some days of LHC operation, and sensitive devices like radios will definitely be among the most affected classes of electronics. In line with the technological solutions for the telecommunication network, the existing leaky-feeder cable was found to be the element that could overcome that limitation, and therefore, the researched localization technologies had this medium as its target.

Among indoor location techniques, those based on the RSSI are of particular interest since they require neither the installation of extra infrastructure hardware nor the allocation of extra spectrum. Besides the simplicity, RSSI methods do not require additional emitters therefore eliminating the risk of interference. Because of these facts, RSSI fingerprinting turned out to be a very interesting approach for localization in the selected scenario. The conducted tests, results, and performance evaluation are presented hereafter.

### 5.4.1 RSSI Fingerprinting over Leaky Coaxial Cable

All RSSI fingerprinting methods are based on a set of common principles. The fundamental point is that during an offline (calibration) phase, a radio map is created, being later used in the online phase to find the position of a set of live samples. In the case of the LHC tunnel, RSSI fingerprinting introduces an attractive challenge. Although this is effectively a 1D localization problem, as only one coordinate axis (along the tunnel) is required, the differences to other indoor scenarios are notable and might introduce serious issues. On the one hand, localization is to be available along all the tunnel's length, that is, nearly 27 km, which besides possessing quite a different shape from that of offices and other spaces, is much beyond the length of any other studied indoors location system or experiment. On the other hand, it is limited to the existing network infrastructure. In the LHC, the only network available throughout the tunnel is GSM, deployed over leaky coaxial cable, providing typically two channels per sector. Because of the low radio frequency attenuation of GSM frequencies in the cable, small variability in received power might lead to large location errors.

To evaluate the applicability and performance of RSSI fingerprinting methods over the leaky-feeder in the LHC tunnel, a series of experiments were conducted. The first series of the experiments characterize the RSSI profile in the tunnel in order to evaluate its usefulness to matching algorithms. In a second series, this evaluation is performed in real-world conditions, where some localization algorithms are tested and benchmarked against a radio map built from the collected RSSI data.

## 5.4.2 Experiments

The experiments were conducted in a sector of the LHC comprising both straight and bending sections. With the help of mobile phones whose API could be accessed from a laptop, thousands of RSSI samples were collected to characterize the network and create the radio map (see Figure 5.3). An online phase with several algorithms was later simulated by using an independent set of samples from the radio map and by using them as live samples to determine location.

### 5.4.2.1 RSSI Profile

In an initial phase, to characterize the network signal strength profile, several types of measurements were carried out. Among them, two are of particular interest:

1. Detailed measurements with fingerprints were taken every 40 m in a section of 600 m and 10 samples per fingerprint. Four such sessions were carried out.
2. Fixed location measurements were taken, where 150 samples of the signal strength were collected at the same position to account for variations depending on the measurement conditions.

**FIGURE 5.3**
Equipment used for data collection.

### 5.4.2.2 Performance of RSSI Fingerprinting

The performance was evaluated in terms of location accuracy with realistic conditions. Besides a dedicated data collection phase, the evaluation phase includes the test of a new algorithm specifically developed for the observed signal characteristics.

*Setup for offline phase:* First, a radio map was built based on several RSSI measurements sessions performed in a slightly curved tunnel sector over a length of 270 m. Fingerprints were taken every 10 m, yielding a total of 28 calibration points. Thirty samples per point were collected simultaneously by two mobile phones, and two complete subsequent sessions were performed. Second, to compare results, wireless LAN (IEEE 802.11) was also evaluated by placing two access points in the ends of the 270 m section of the tunnel (see Figure 5.4). RSSI fingerprints were collected at identical positions as GSM, that is, every 10 m, each counting 100 samples.

Samples were preprocessed per point and network technology to improve performance during the online phase while considerably reducing the database size. Several statistical parameters describing each fingerprint distribution were calculated:

- Minimum and maximum
- Quartiles
- Histogram, fenced at 1.5*inter-quartile range (IQR)
- Histogram high and low fence

By saving the histogram fenced according to the IQR, one significantly reduces the presence of outliers while efficiently storing the samples' RSSI information. Moreover, from this representation one can very easily obtain other statistical measures, like the median (directly from the second quartile).

*Evaluating RSSI fingerprinting methods:* In the online phase, the location algorithms were evaluated, and, to assess their performance with a reasonable

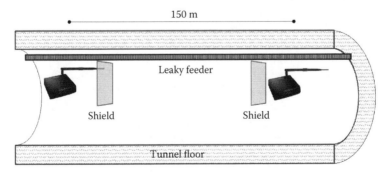

**FIGURE 5.4**
WLAN experiment setup, promoting signal injection in the leaky-feeder cable.

confidence level, they were tested with independent sets of samples taken in addition to those used to create the radio map. Ten samples were used for each online fingerprint. In the current configuration of two measurement sessions with two receiver units, this yields four fingerprints for each of the 28 points. To simplify and automate the execution of the tests, a software framework for localization was implemented. The software allows for transparent handling of different data sets, postprocessing and comparative performance analysis of the different implemented location-finding algorithms.

The tested location-matching algorithms are based on the KNN (K-Nearest Neighbors) method, having some decisions and parameters changed to meet the framework of the experiment. Among those parameters, the median was used as the base statistical function. For the absolute RSSI value comparison, a KNN variant (Pereira, Theis, Moreira, & Ricardo, 2012) was developed and applied. This employs 2-norm (Euclidean) distances and weights each channel contribution by the probability of the fingerprint to be compliant with the normally approximated RSSI distribution of that point's samples. The method showed to be appropriate for situations where large variances in RSSI are likely to happen even at each single point.

A second algorithm was developed that takes advantage of the RSSI difference among different channels, here denominated inter-channel-RSSI-differential (ICRD). This is motivated by the fact that small changes in position might affect all channels similarly, but their relative power will remain constant and might reflect the position RSSI profile more accurately. In the case of channels propagating in the same direction, it has been observed that they actually correlate very closely, and therefore these channels, denominated same-propagation-direction ICRD (SPD-ICRD), are considered independently.

Basically, after an initial step of identifying the channels common with the online fingerprint, the algorithm acting on the ICRD map works by calculating the differential between the channels in a circular order. For instance, if there are four common channels, for example, channels 1, 2, 3, and 4, then only the differentials 1-2, 2-3, 3-4, 4-1 are used. With this technique, the algorithm avoids the potentially large number of comparisons resulting from all the possible channel combinations. Moreover, because of the circular link, absolute RSSI changes in any channel will always affect only two differentials independently of the total number of channels found in the radio map. $D$ is the ICRD, $SD$ is the SPD-ICRD, $N$ is the number of channels, $k$ is the SPD-ICRD weight, $i$ is the online location index, and $p$ is the radio-map location index.

According to Equation 1, this algorithm takes all differentials (ICRDs), including the SPD, and attributes them the same weight. To take advantage of the high-correlation properties with the same direction channels, this differential is given a higher weight ($k$), but since it had been considered once in the summation part of the equation, the weight becomes ($k - 1$). Values in the range of 3 to 10 were tested, and experimental results suggest that higher values are better suited for radio maps with higher resolution.

Taking the ICRD metric into account for a localization algorithm, in addition to absolute RSSI, led to the development of a hybrid method. Furthermore, fingerprints from GSM and WLAN were fed both independently and combined to the algorithms, resulting in a total of six tests:

1. Weighted KNN with GSM
2. Hybrid (weighted and ICRD) KNN with GSM
3. Weighted KNN with WLAN
4. Hybrid KNN with WLAN
5. Multitechnology weighted KNN (GSM and WLAN)
6. Multitechnology hybrid KNN

### 5.4.3 Results

According to the specified experiments, this section presents two groups of results: in the first group, a characterization of the GSM RSSI signal in the tunnel, and in the second group, an evaluation of localization algorithms using the collected fingerprints.

#### 5.4.3.1 Profile of GSM RSSI over Leaky-Feeder

As described in Section 5.4.2.1, the RSSI profile is evaluated for both position changes and also depending on measurement conditions. A third test presented here shows the RSSI evolution according to distance in a different tunnel section and compares GSM alongside with wireless LAN.

*RSSI according to position:* As we go along the tunnel, we notice that the received signal strength can change significantly. In Figure 5.5 we present the mean and standard deviation of the RSSI values for each detailed measurement as a function of the distance. We can clearly see two dominant GSM channels and traces of several channels from nearby tunnel sections. In the plot we can also identify two distinct areas, specifically before and after measurement location 6. In fact, the leaky-feeder is injected with a new GSM channel at this location. Until then, the new frequency experiences high attenuation as it propagates through air.

After that (to the right of measurement location 6), the propagation occurs normally in the cable, and therefore, the attenuation is much lower. In this region, approximating the RSSI evolution to a linear function, we obtained attenuation factors of 3.9 dB/km and 4.5 dB/km, slightly larger than the attenuation of the cable. Despite the fact that the cable's longitudinal attenuation introduces measurable signal changes, the variations in the RSSI values tend to obfuscate them, dramatically reducing the accuracy of the estimation. These effects are generally caused by the volatility of the measurement

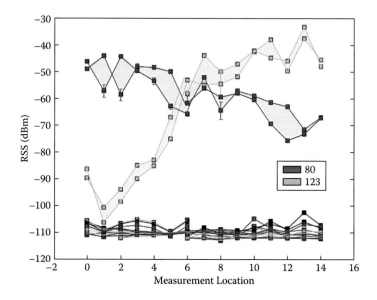

**FIGURE 5.5**
RSSI for consecutive measurement points. Each line color represents a different frequency, error bars represent standard deviation, and shaded areas are delimited by RSSI values of the same frequency obtained in different measurements.

conditions but are also due to the influence of *multipath* propagation in the tunnel and *multicoupling* in the leaky-feeder itself (Weber et al., 2010).

*RSSI dependence on measurement conditions:* Although the four detailed measurement sessions were performed in rather similar conditions, the observed RSSI values at each point exhibit divergent behaviors. It was rather surprising to find almost null variance in two consecutive measurements while their average differs by more than 10 dB (e.g., measurement point 11 in Figure 5.5). This fact motivated the stationary measurements in where we tested three slightly different conditions:

- *Optimal:* We ensured no one was close to the equipment during the measurement process, by at least 30 meters.
- *Suboptimal:* At least one person was standing beside the equipment during the measurement.
- *Realistic:* One person was holding the equipment and slightly moving it during the measurement.

Figure 5.6 shows the RSSI evolution for the measurements under the different conditions. The more adverse the conditions the higher the variations, and the averaged RSSI values tend to drop. Such observations are clearly confirmed by their respective histograms (see Figure 5.7), where one clearly notices a larger spread of the distribution and a small shift to the left.

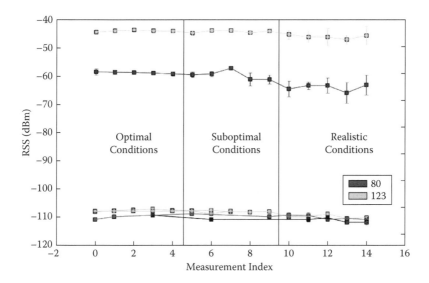

**FIGURE 5.6**

RSSI dependence on measurement conditions. Each group was taken with slightly different conditions: (a) optimal, (b) suboptimal, or (c) realistic.

The results show that the environmental characteristics of this underground tunnel, under optimal conditions, cause very few signal fluctuations, which is quite favorable for fingerprinting methods. However, for scenarios requiring some manual handling of the equipment, as will be the case for radiation surveys, one must account for significant signal variations, which will inevitably affect the accuracy.

*Comparison of RSSI profiles between GSM and WLAN:* The tests performed to compare the performance of RSSI fingerprinting between GSM and WLAN provide immediate insight of the respective power profiles by direct inspection of their radio maps.

Figure 5.8 shows the RSSI evolution for the GSM network along the tunnel section. One can clearly see the two dominant channels (ID 123 getting stronger and ID 97 attenuating) and a third weaker channel also getting stronger. In turn, the WLAN plot (see Figure 5.9) reveals a considerably higher degree of attenuation relative to GSM, on the order of 14 dB/100 m. This factor generally favors the performance of fingerprinting-based methods. Such difference in attenuation comes as a direct consequence of the higher frequencies used. While the leaky-feeder has a 3.16 dB/100 m attenuation figure for 900 MHz, the attenuation at 2.4 GHz should be higher than 10 dB/100 m (the closest frequency for which there is specification is 1,900 MHz, exhibiting 8.52 dB/100 m).

### 5.4.3.2 Evaluation of Location Accuracy

Various methods were implemented to evaluate the effectiveness of each algorithm in determining location. At first, a scoring method assessed the

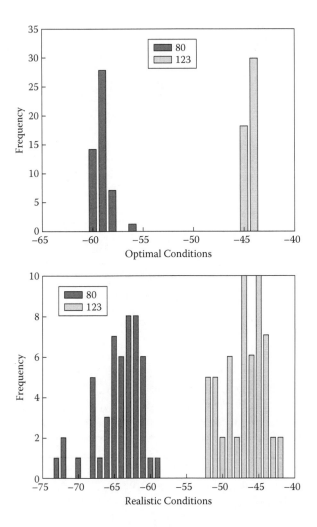

**FIGURE 5.7**
RSSI histograms for two distinct measurement conditions.

quality of the location estimate according to the confidence levels obtained by the algorithm with respect to the true location. This method gives a good insight into the quality of the algorithm and the radio map, but it cannot be used to estimate positions and accuracy levels. Therefore, performance was also evaluated in terms of the actual accuracy achieved with KNN, calculated for K = 1, 3, 5, and 7.

With GSM it is noticeable that the performance of all methods gets better as more locations (K) are considered by KNN (see Figure 5.10). This might be due to the fact that with such little differences in absolute value, finding the exact match of a location is very difficult, and averaging the best guesses weighted by their degree of confidence indeed becomes a favorable option.

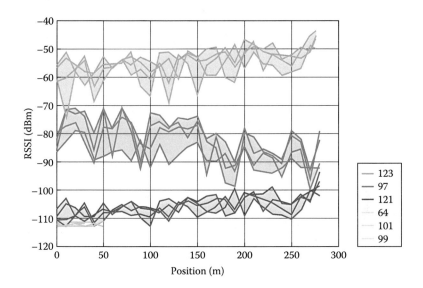

**FIGURE 5.8**
GSM RSSI evolution of different channels, measured in four sessions.

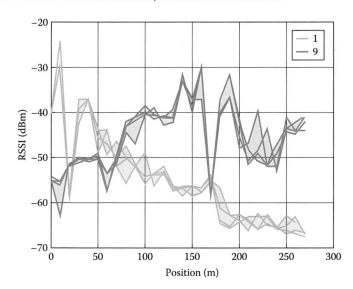

**FIGURE 5.9**
WLAN RSSI evolution of different channels, measured in four sessions.

Evaluating the methods performance for WLAN (see Figure 5.11), one can clearly observe the significantly better performance of the absolute RSSI matching method than the differential one. Nevertheless, as expected, the absence of multiple ICRD channels has a very negative impact, and the average location error using the ICRD method is above 30 m. Yet, this value is

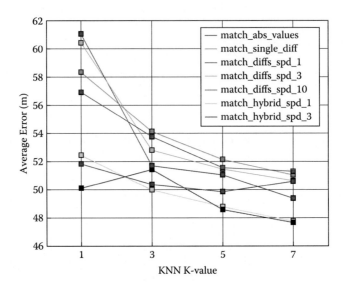

**FIGURE 5.10**
Performance of the various matching algorithms with KNN, evaluated for GSM for K = 1, 3, 5, and 7.

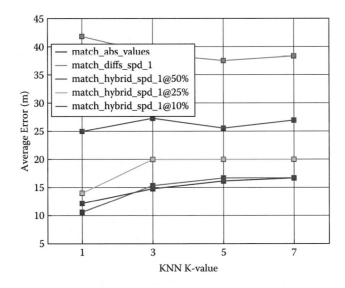

**FIGURE 5.11**
Performance of the various matching algorithms with KNN, evaluated for WLAN for K = 1, 3, 5, and 7.

still lower than the best method working over GSM. This comparison clearly demonstrates that the performance of location fingerprinting methods is highly dependent on the underlying signal propagation characteristics. In addition, in this case, the best performance with KNN is obtained for the lower values of K. This result complies with the explanation that the higher the signal attenuation along the tunnel, the more accurate are the first guesses of the methods. In this case, for K = 1, the absolute RSSI method yields an average distance error of 12.1 m.

To assess the performance achieved by combining multiple technologies, several possible configurations based on the previous algorithms were tested. A summary of the methods settings is presented in Table 5.1, and both simple and hybrid and multitechnology algorithms are considered.

The results are shown in Figure 5.12 and Figure 5.13, which provide a good overall performance summary. The performances of the absolute RSSI method for GSM (gsm_abs_values) and WLAN (wlan_abs_values) are included for base comparison, as both are then combined by the multitechnology method (gsm-abs+wlan-abs).

The difference in the accuracy among the different radio technologies is obvious. With GSM the best accuracy is nearly 90 m, while with WLAN the accuracy is 20 m (at 90% confidence). When combining both technologies giving 50%–50% weights, the performance of this multitechnology method largely exceeds that from GSM alone and approaches the one from WLAN. By giving 10% weight to GSM, gsm-abs+wlan-abs (10%), the global performance improves and achieves the same level as WLAN itself.

**TABLE 5.1**

Parameters of the Compared Algorithms

| Name | Technology (Weight) | Matching Algorithm | ICRD Weight |
|---|---|---|---|
| gsm-abs_values | GSM (100%) | match_abs_values | 0 |
| gsm-hybrid-spd3 | GSM (100%) | match_hybrid | 50% (SPD = 3) |
| wlan-abs_values | WLAN (100%) | match_abs_values | 0 |
| wlan-hybrid@10% | WLAN (100%) | match_hybrid | 10% |
| gsm-abs+wlan-abs (50%) | GSM (50%) | match_abs_values | N/A |
| | WLAN (50%) | match_abs_values | N/A |
| gsm-abs+wlan-abs (10%) | GSM (10%) | match_abs_values | N/A |
| | WLAN (90%) | match_abs_values | N/A |
| gsm-hybrid-spd3+wlan-ybrid@10% (10%) | GSM (10%) | match_hybrid | 50% (SPD = 3) |
| | WLAN (90%) | match_hybrid | 10% |

*Note:* ICRD = inter-channel-RSSI-differential, GSM = , WLAN = ,
SPD = same-propagation-direction.

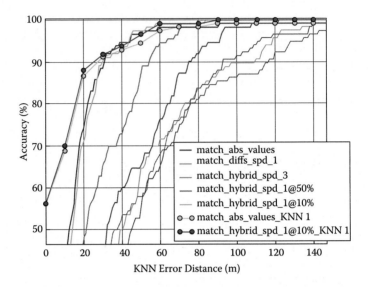

**FIGURE 5.12**
Accuracy achieved by the various methods with KNN (K = 7, otherwise indicated). Thin lines are relative to GSM, while bold lines are relative to WLAN. As can be seen the WLAN results clearly outperform the ones obtained with GSM, independently of the matching algorithm.

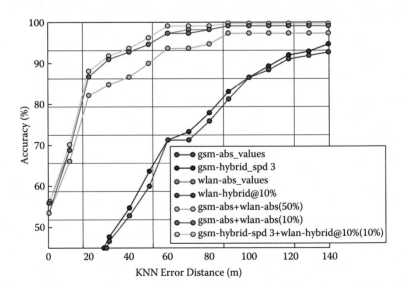

**FIGURE 5.13**
Accuracy comparison among various methods, simple, hybrid, and multitechnology, evaluated with KNN (K = 1). The methods' parameters are explained in Table 5.1. Please note that red and yellow lines are totally coincidental, while green and purple lines also coincide except at one point.

## 5.5 Conclusions and Open Challenges

In this chapter, an overview of the challenges and opportunities for positioning and localization in underground environments was presented. A particular case study was described, showing the localization experiments conducted in the LHC tunnel by leveraging the preexisting GSM infrastructure using fingerprinting techniques. We realized that although the GSM RSSI remained impressively constant under optimal measurement conditions, significant differences arise among different measurement sessions. This could be because the equipment located in the accelerator tunnel changes its operation modes (e.g., magnets being powered) or because the tunnel configuration promotes fast-fading phenomena.

When testing localization algorithms, we explored two KNN approaches: a modified weighted KNN working over the absolute channels' RSSI, and a hybrid algorithm taking additional advantage of the channels' RSSI differential. The latter method achieved slightly higher location accuracy, and the fact that the RSSI differential is more resilient to network changes helps reduce the need for recalibration. Given the simplicity of the methods, one could expect that the results would still be conservative. However, when we simulated an ideal KNN over the current data (Pereira et al., 2012), it was shown that it could only yield an improvement by 20% to 30% in the accuracy.

By exploring the signal strength of the WLAN network, we can obtain much higher accuracy levels, in the order of 20 m at 88% confidence. Combining results from both the GSM and WLAN technologies performs relatively well, and although significant improvement on the overall result could not be observed, the achieved accuracy is as high as the best underlying technology alone. This might be because there is a significant discrepancy between the achieved accuracy levels for WLAN and GSM. In addition, performance comparison plots between WLAN and GSM also show evidence that the KNN parameter K should be chosen according to the attenuation factor of the network, where higher values of K favor only radio maps having little attenuation between adjacent points. The current WLAN setup, specifically deployed for comparison reasons, was limited to two access points. Larger deployments of WLAN and eventually other network technologies, like TETRA, might also help improve accuracy figures.

Although localization based on fingerprinting techniques with received signal strength performs well, it yields accuracy levels in the order of 20 m for WLAN and 90 m for GSM (confidence of 90%), which is still not sufficient for the most demanding applications envisaged for the system. To increase the accuracy up to the envisaged levels, techniques based on time of flight (radio frequency wave propagation delay) that could meet the tunnels' restrictions and specificities are currently being investigated. By using frequencies in the VHF band (2 m wavelength), the technology aims at achieving 1-meter-level accuracy, and by propagating the signal over the leaky-feeder cable, full tunnel coverage is expected to be achieved with a small number of units. To allow for fast prototyping and custom deployment of the methods at a relatively low cost, programmable software-defined radio devices are being considered for implementation.

# References

Bahl, P., & Padmanabhan, V. N. (2000). RADAR: An RF-based in-building user location and tracking system. In *Proceedings of IEEE Infocom* (pp. 775–784). New York, NY: IEEE.

Chon, H. D., Jun, S., Jung, H., & An, S. W. (2004). Using RFID for accurate positioning. *Journal of Global Positioning Systems, 3*(1–2), 32–39.

Dayekh, S., Affes, S., Kandil, N., & Nerguizian, C. (2011). Cooperative geo-location in underground mines: A novel fingerprint positioning technique exploiting spatio-temporal diversity. In *2011 IEEE 22nd international symposium on personal, indoor and mobile radio communications* (pp. 1319–1324). New York, NY: IEEE.

Engelbrecht, J., Collmann, R., Birkel, U., & Weber, M. (2010). Methodical leaky feeder design for indoor positioning considering multipath environments. In *Radio and wireless symposium (RWS), 2010 IEEE* (pp. 164–167). New York, NY: IEEE.

Engelbrecht, J., Collmann, R., Birkel, U., & Weber, M. (2011). First results of a leaky coaxial cable prototype for indoor positioning. In *Wireless telecommunications symposium (WTS 2011)* (pp. 1–5). New York, NY: IEEE.

Engelbrecht, J., Forster, G., Michler, O., & Collmann, R. (2012). Positioning estimation in public transport systems by leaky coaxial cables. In *9th workshop on positioning navigation and communication (WPNC)* (pp. 175–179). New York, NY: IEEE.

Fairfield, N., Kantor, G., & Wettergreen, D. (2007). Real-time SLAM with octree evidence grids for exploration in underwater tunnels. *Journal of Field Robotics, 24,* 3–21.

Fink, A., Beikirch, H., Voss, M., & Schroder, C. (2010). RSSI-based indoor positioning using diversity and inertial navigation. In *2010 international conference on indoor positioning and indoor navigation (IPIN)* (Vol. 1, No. 7, pp. 15–17). New York, NY: IEEE.

Mautz, R. (2012). Indoor Positioning Technologies Thesis. ETH Zurich.

Michalson, W. R., & Pogri, I. F. (2000). Assessing the accuracy of underground positioning, using pseudolites. ION GPS-2000. Technical Meeting of Institution of Navigation.

Misra, P. (2009). Underground mine communication and tracking systems: A survey, UNSW-CSE-TR-0910. University of New South Wales.

Nakamura, M., Takagi, H., Terashima, J., Einaga, K., Nishikawa, T., Moriyama, N., & Wasaki, K. (2010). Development of a simple multiple-position identifying system with a long range multiband leaky coaxial cable for rescue operations in tunnels or passages in underground facilities. In *Microwave conference proceedings (APMC), 2010 Asia-Pacific* (pp. 163–166). New York, NY: IEEE.

Pereira, F., Theis, C., Moreira, A., & Ricardo, M. (2012). Performance and limits of KNN-based positioning methods for GSM networks over leaky feeder in underground tunnels. *Journal of Location Based Services, 6*(2), 117–133.

Phithakkitnukoon, S., Wolf, M. I., Offenhuber, D., Lee, D., Biderman, A., & Ratti, C. (2013). Tracking trash. *Pervasive Computing, IEEE, 12*(2), 38–48.

Schneider, O. (2010). Requirements for positioning and navigation in underground constructions. In *2010 international conference on indoor positioning and indoor navigation (IPIN)* (Vol. 1, No. 4, pp. 15–17). New York, NY: IEEE.

Want, R., Hopper, H., Falcão, V., & Gibbons, J. (1992). The active badge location system. *ACM Transactions on Information Systems, 10*(1), 91–102.

Weber, M., Birkel, U., Collmann, R., & Engelbrecht, J. (2010). Comparison of various methods for indoor RF fingerprinting using leaky feeder cable. In *7th workshop on positioning navigation and communication (WPNC)* (pp. 291–298). New York, NY: IEEE.

Weber, M., Birkel, U., Collmann, R., & Engelbrech, J. (2011). Wireless indoor positioning: Localization improvements with a leaky coaxial cable prototype. Guimareas. Pp. 21–23.

# 6

# *Map-Aided Indoor Navigation*

Susanna Kaiser

Mohammed Khider

Maria Garcia Puyol

Luigi Bruno

Patrick Robertson

## CONTENTS

*Abstract:* Maps representing aspects of an environment that affect pedestrian motion can be very informative sources of data in indoor localization. Their proper representation and usage are mandatory to fully leverage their potential. In this chapter, we show how probabilistic representations facilitate accuracy and availability of position estimates even in the absence of usable satellite navigation signals or similar forms of localization signals. We will show that maps may effectively substitute infrastructure, such as active or passive (RFID-type) radio beacons, when their information is properly used in combination with dynamic models of movement and some form of motion estimate such as pedestrian dead reckoning (PDR). This chapter aims at illuminating the details of how to generate, represent, and use probabilistic maps for indoor localization. While this discussion applies to a wide range of sensors, we will focus on showing how maps are essential in achieving long-term stability in combination with inertial sensors. We begin by motivating why the use of a probabilistic map of human motion is a natural way of incorporating building information into a sequential Bayesian filtering framework. This stands in contrast to the oft-used ad hoc solution, which is to use a floor plan as a "kill or live" weighting function in a particle filter (PF), driven by some form of PDR such as foot-mounted inertial-sensors-based PDR. We show how the latter method can fail catastrophically and how a probabilistic map formulation addresses these problems. We present a number of ways of how to obtain such maps for real-world applications. The first is based on knowledge of the building layout and applies a diffusion algorithm to compute an estimate of the probability distribution of the motion direction of a pedestrian at each point in the building. Second, we compare these maps with those obtained using simultaneous localization and mapping (SLAM) by applying FootSLAM, which requires no sensors other than a source of dead reckoning. The map concept can be further extended in order to include features that are relevant to radio-based localization techniques, like transmitter positions and a model for radio propagation or, eventually, a database of fingerprints. The influence of the different kinds of maps on positioning accuracy is discussed in detail, and the maps are compared to each other by means of metrics derived from information theory.

## 6.1 Introduction

The ability to navigate, that is, to determine one's location and find a route to a desired destination, has always been desirable for human beings. Nowadays, this capability is prolific, due to global navigation satellite systems (GNSSs) and corresponding low-cost receivers available in a wide range of mobile devices, such as smartphones. Market studies predict as many as 1.2 billion GNSS-enabled mobile devices by 2015 (Galileo Ready Advanced Mass Market Receiver, 2009). In addition to this strong and increasing presence in mass-market applications, satellite navigation has become essential for professional applications ranging from logistics to security. This success demonstrates both the demand for navigation and the technological suitability of GNSSs to deliver. Our work is motivated not only by this success but also by the significant challenges posed by radio propagation for GNSS-based localization in indoor environments. GNSS and other radio signals are affected by blocking and multipath propagation, which significantly deteriorate the positioning accuracy and availability in indoor environments and urban canyons. Important contributions have been made by the navigation research community to address these problems both with and without a preinstalled infrastructure. However, we believe that GNSS-only solutions will be inadequate in providing truly ubiquitous coverage for highly accurate navigation applications for pedestrians in indoor environments. This is because the accuracy requirements indoors can be seen to be much higher than outdoors; a universal way-finding application, for instance, would need floor-level accuracy and horizontal accuracy of around 1 meter.

The overall goal of this chapter is to provide an overview of how probabilistic maps can help obtain long-term stability in indoor navigation algorithms. The term *map* implies not only the use of available building plans but also the use of a map in a probabilistic way including learning the probabilistic map during the walk. This chapter is organized as follows. In the remainder of this section, we give an overview of existing infrastructure-based and infrastructure-free indoor navigation techniques (Sections 6.1.1 and 6.1.2), describe multisensor approaches (Section 6.1.3), discuss the possibility of learning probabilistic maps during the walk (Section 6.1.4), and give an overview of how human motion can be considered in the concept of probabilistic maps (Section 6.1.5). Section 6.2 illuminates the generation of a probabilistic map from floor plans and satellite images. Sections 6.3 and 6.4 describe different ways of learning probabilistic maps during the walk using SLAM. Section 6.5 provides experimental results in terms of quantifying maps with an entropy metric and calculating the position accuracy. Section 6.6 outlines the main conclusions.

### 6.1.1 Infrastructure-Based Approaches

Some researchers have focused their work on standalone positioning sensors, such as the work on Wi-Fi positioning in Kotanen, Hannikainen, Leppakoski, and Hamalainen (2003), radio-frequency identification (RFID) positioning in Wendlandt, Robertson, Khider, Angermann, and Suchaya (2007), and ultrawideband (UWB) positioning in Pietrzyk and von der Grün (2010). The dividing line between the infrastructure-free and infrastructure-based approaches is actually blurred. If available Wi-Fi base stations are to be used, then no additional positioning infrastructure would be needed. For the purpose of this chapter, we will treat the use of Wi-Fi base stations as infrastructure-based.

The most promising results for pedestrian localization allowing quasi-ubiquitous coverage based on an available infrastructure have been achieved using systems that rely on the measurement of received signal strength (RSS) from Wi-Fi access points (APs) such as those in Ibach, Stantchev, Lederer, White, Herbst, and Kunze (2005), Anon. (2006), Bahl and Padmanabhan (2000), and Kotanen, Hannikainen, Leppakoski, and Hamalainen (2003). Two different approaches are widely used as the physical basis for Wi-Fi signal-strength-based localization (Liu, Darabi, Banerjee, Hamalainen, and Liu, 2007):

- A *path loss model approach*, typically dependent on the indoor propagation models established by the COST 231 standards (COST Action 231, 1999), estimates the attenuation of signal strength over distance in space from the known location of APs and their known transmit power and uses trilateration to estimate the user position, as in Kotanen, Hannikainen, Leppakoski, and Hamalainen (2003).

- An *empirical approach* relies on prerecorded calibration data in order to generate an RSS map of the environment (Zarimpas, Honary, Lund, Tanriover, & Thanopoulos, 2005), widely known as "fingerprinting."

The disadvantage of these positioning techniques is, of course, their dependence on infrastructure. In addition to maintenance issues, the availability of the infrastructure might not be given in some important use cases such as those arising in emergency situations. Note that structure-based approaches require the knowledge of the RSS map or of the position of the APs, which can also be considered as a form of map, and these maps are a subject of this chapter.

### 6.1.2 Infrastructure-Free Approaches

To avoid being infrastructure dependent, other techniques focus on sensors that do not require additional local infrastructure, such as long-term evolution mobile radio signals (Staudinger, Klein, & Sand, 2011) or sensors that the pedestrian carries while walking, such as electronic compasses, barometers, and inertial measurement units (IMUs). In the case of mobile radio signals, it can be argued that their provisioning is usually outside the domain of the building operator and hence considered to be infrastructure-free, like GNSSs. However, mobile radio-based approaches might also be subject to vulnerabilities in events like natural disasters.

IMUs have shown very promising results as infrastructure-free positioning sensors and have been widely investigated by the navigation community accordingly to provide position and orientation information (Harle, 2013). An IMU is an electronic device that provides inertial measurements, usually using a combination of accelerometers and gyroscopes. Distance traveled from an initial position and attitude in space can in principle be calculated from these measurements. However, commercially available low-cost IMUs exhibit severe drift and noise and hence provide only very short-term stability (Woodman, 2007). If the sensor is placed on the foot of a pedestrian, zero-velocity updates (ZUPTs) (Foxlin, 2005) and other corrections during the stance phase of the human stride can limit the temporal error growth from cubic to linear. However, positioning based only on ZUPT-assisted IMUs suffers from the following limitations: it lacks a link to a global reference; additional errors may arise because of insufficiently modeled sensor effects caused by disturbances, false-detected ZUPTs, and unexpected human-movement behaviors; and the cumulative error growth will still limit the applicability to only very few application domains without any other knowledge or assumptions. As a way of addressing these limitations, the use of a priori knowledge of floor plans has been proposed in Beauregard, Widyawan, and Klepal (2008), Krach and Robertson (2008), and Woodman and Harle (2013). We will address these approaches in Section 6.2.

The error of inertial positioning for pedestrians can be quantified as one component affecting the estimated distance traveled and one affecting the estimated angle of a displacement, such as a human step (Nilsson, Skog, & Händel, 2010; Skog, Nilsson, & Händel, 2010; Zampella, Khider, Robertson, & Jiménez, 2012). Both error components of course affect the resulting position estimate, but an angular error event at any given time will lead to a greater resulting position error if the pedestrian moves further away over time. Fortunately, the structure of buildings imposes constraints on human motion, and the aforementioned approaches use knowledge of these structural constraints to curtail the error growth while a pedestrian moves in the structured environment.

### 6.1.3 Multisensor Approaches

A state-of-the-art approach to achieving more accurate positioning is to combine the outputs of two or more usually heterogeneous positioning sensors, which is referred to as multisensor positioning (Ristic, Arulampalam, & Gordon, 2004). When the sensors involved are heterogeneous (of different types and characteristics) and subject to noise, combining their measurements can improve the accuracy and reliability of the estimated position. Sequential Bayesian algorithms are based on fundamental, probabilistic estimation theory. They have shown the very promising results due to their ability to incorporate a priori knowledge about pedestrian dynamics, sensor characteristics, and the environment into the estimation process, as they interpret sensor outputs using probability densities and take into account the difference in accuracy associated with using different sensors. Moreover, they allow one to provide probabilistic location estimates rather than just point estimates. These probabilistic estimates encode the confidence in the estimates and can be used by client applications.

To provide some examples of sensor-fusion approaches, Chiu and O'Keefe (2008) illustrated the benefit of the integration of UWB and GPS, and Weyn and Schrooyen (2008) demonstrated a concept for Wi-Fi-assisted GPS positioning. A Wi-Fi fingerprinting approach fused with inertial sensors is detailed in Frank, Krach, Catterall, and Robertson (2009). Further examples of IMU-based multisensor navigation systems can be found in Soloviev and Miller (2010) and Klingbeil, Romanovas, Schneider, Traechtler, and Manoli (2010).

The range of possible sensors is vast; recently, Frassl, Angermann, Lichtenstern, Robertson, Julian, and Doniec (2013) showed how to use a map of the local distortions of the magnetic field in a building in combination with odometry; that is, estimation of the change in position over time from data of moving sensors to improve indoor localization of robots and pedestrians.

### 6.1.4 When Maps Are Unknown

All infrastructure-free approaches mentioned so far rely on some prior knowledge of the environment, such as the position of the APs (Kotanen et al., 2003), the radio map (Zarimpas et al., 2005), the magnetic field map (Frassl et al., 2013), or the floor plan (Krach & Robertson, 2008). Therefore, they are critical to the rollout, cost-efficiency, reliability, and accuracy of positioning services.

When maps are partially or completely unavailable, one can sometimes resort to the SLAM principle (Durrant-Whyte & Bailey, 2006) to generate a suitable map representation without a dedicated mapping phase that needs knowledge of location. Although SLAM has its origins in the field of robotics, a number of approaches have proposed SLAM for pedestrians to generate a radio map (Bruno & Robertson, 2011; Ferris, Fox, & Lawrence, 2007), a map of the local distortions of the magnetic field (Robertson et al., 2013), a probabilistic map of human motion (Robertson, Angermann, & Krach, 2009), a map of location-related actions (Hardegger, Roggen, Mazilu, & Troster, 2011), or a map of all available signals (Mirowski, Ho, Yi, & MacDonald, 2013). In SLAM, the estimation problem is formulated as one comprising both the location of an agent and an environment that, when appropriately sensed, is informative of the location. There are many different potential ways to manage and align the map-gathering process when performing SLAM. A simple cleaning robot might perform SLAM in order to position itself in a building while cleaning—and discard the map when done. In the context of pedestrian localization, SLAM may be performed on data collected from many people to generate maps that are then used by (other) pedestrians (Robertson, Garcia Puyol, & Angermann, 2011; Shen, Chen, Zhang, Moscibroda, & Zhang, 2013). It is important to note that SLAM may benefit greatly from available maps, even if these are not complete or as accurate as the desired map. This was shown in Kaiser et al. (2012) and will be discussed in Section 6.5.

### 6.1.5 Human Motion and Probabilistic Maps

#### 6.1.5.1 Human Motion in Constraint Environments

The most important overall contribution of this chapter is to show how the interplay between human motion and the buildings humans inhabit can be used to provide accurate infrastructure-free positioning with low-cost sensors. In indoor environments, the presence of walls, doors, and furniture constrains and channels our motion as pedestrians. Some areas, such as doorways and narrow corridors, force us to walk in almost the same fashion when visiting these areas. Other areas, such as large and empty rooms, allow more freedom. In addition to these physical aspects of the environment, there are countless characteristics that interact in a very complex and individual fashion with our own biological sensors and cognitive processing to guide our motion. A sign to an exit might be relevant to our future motion if we are looking for the way out of a building, for example, but might be irrelevant to somebody else with different intentions. We can reasonably assume that these influences on our motion are subjective, time variant, and dependent on a large set of additional factors, which we expect to be impossible to quantify exactly. However, despite these complexities, we believe that we can derive useful maps simply by observing the actual motion of humans. If we were able to observe the individual trajectories of a large number of people at different times while they walk in an area under study, then we would make the following argument of having the basis for an anonymous statistical model that would aggregate all observed motion patterns. A model of this nature was proposed by Krumm (2008) for predicting a driver's near-term future route based on his or her past trajectory and observations of that driver's history of trajectories. At any particular location and, conditioned on the history of a person's trajectory, we would be able to use this model to estimate the probability of each possible future motion pattern. We would expect that the motion history in the immediate past would have more influence than the motion pattern lying further back in time. Such a model could be represented for instance as an $n$th order Markov process, as proposed in Krumm (2008).

Since we are interested in obtaining maps that reflect the structure of the environment, we can limit the unbounded error growth inherent to inertial navigation systems. In this case, the physical constraints (or lack thereof) within an environment could be estimated by observing trajectories from many people, since there are not many individuals who are able to walk through walls. A door that is never used is structurally and statistically identical to a wall at its place. As we will explain in Section 6.3, FootSLAM is an approach that uses SLAM to estimate a probabilistic map of human motion (Robertson et al., 2009), capturing the constraints of human motion as the probability distribution, at a particularly small region of space, of a person's next step. A wall would be reflected in making a step in its direction very small or zero, once one is close to the wall. FootSLAM conducted on data from real people can also capture some of the more subtle influences of the environment on human motion, such as our tendency to walk along straight lines much of the time and in the way we negotiate corridors. The applicability of these models from one person to another or from a group of people to an individual is a topic for further research.

In the following sections we will look at ways of computing models for human motion based on structural knowledge of a building, rather than observing people as they walk in that building.

### 6.1.5.2 Realistic Motion Models

As mentioned earlier, much work has been performed to use known constraints, such as floor plans, as a source of information in a sequential Bayesian estimation process of a pedestrian's trajectory given a source of noisy odometry, assuming a suitable "map" to be available. Early work (Beauregard et al., 2008; Krach & Robertson, 2008; Woodman & Harle, 2008) considered walls as obstacles that constrained motion; that is, walls would not allow a pedestrian to cross them. Walls were assumed not, however, to exert any other influence on human motion. Clearly this is not the case; people tend not to walk right up to a wall. Walls guide our motion even when they are at some distance. Applying a simplistic constraint-only model in sequential estimation processes can yield good localization performance in many situations. However, as explained in more detail in Kaiser, Khider, and Robertson (2011), there are situations where this simple model can fail. On the one hand, the convergence time may be long if the starting position is unknown even in cases with a known floor plan. On the other hand, it might fail when the posterior follows a *multimodal* distribution, which is often true in multisensor fusion approaches. Noisy, erroneous, and heterogeneous sensor measurements, as well as a particular building layout, are some of the reasons for such multimodal posterior distributions. The reasons for this susceptibility to failure are somewhat complex. There are two valid and related explanations. The first is that a constraint-only model is mathematically inaccurate, in that a state transition model of a Bayesian estimator needs to accurately represent probabilistic human motion at any point in space. The second is to look at the reason for failure in multimodal estimation situations when using approximate Bayesian estimators like PFs. There are two opposed approaches to PF-based position estimators, whereas many of the standard PFs use an importance density that is based on the transition prior and use likelihoods based on measurements to weight the particles; in our approach we use odometry measurements to compute a proposal function that represents the estimate of the next step. The weighting is then done using additional probabilistic constraints on human motion. If we choose to simply weight in a binary fashion that assigns weight zero to particles that have crossed a wall, we typically observe a tendency for the estimator to unfairly favor hypotheses that are far from constraints like walls. The link to the previous mathematic explanation is that the evolution of a hypothesis that is close to walls should be rewarded more strongly by the state transition function than one that is far from walls. One example of a multimodal scenario is discussed next (Kaiser et al., 2012). Let us consider a scenario with a large corridor split into two almost parallel corridors leading to two different parts of a building, denoted here as part A and part B (the Y-scenario; see Figure 6.1). The difference between the two parts of the building is in their wall arrangement: part A consists of a narrow corridor forming a loop and has several walls, whereas part B has fewer walls. The true path leads to part A and is given in green. The trajectories in blue show the paths followed by the particles, and their thicknesses depict the contribution of these particles to the posterior density. After entering parts A and B, the particles in area A will suffer a stronger reduction in their population, as they will cross more walls. Particles that enter area B will suffer less reduction and will dominate the posterior over time, leading to a significant position error.

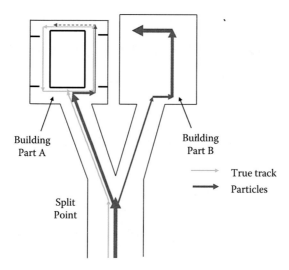

**FIGURE 6.1**

The Y-scenario (from Kaiser et al., 2012; www.tandfonline.com): particles will be divided into two groups if the heading difference between the two paths is small. Because of the denser wall constraints in part A, the particles will have a higher rate of loss and the wrong group will survive in part B.

One can deliberately provoke other scenarios leading to multimodality. Further examples are also given in Kaiser et al. (2012). Kaiser and Khider et al. (2012) argued that an appropriate model incorporating known walls and other constraints should encode the probabilistic influence of the physical environment on human motion. Two different approaches (Kaiser et al., 2011; Khider, Kaiser, & Robertson, 2012) to estimate such a model have been proposed, and both of these are based on gas diffusion models in buildings as a basis for estimating how people might walk in the vicinity of walls and with a given intended destination, in the case of the destination-based model. As a result, these models are able to handle multimodal situations.

Motion models (MMs) are formerly used to probabilistically represent the stochastic nature of pedestrian movement in both indoor and outdoor environments. MMs have mainly been applied to planning tasks, such as estimating pedestrian flows when designing railway stations or airports, or the optimization of evacuation procedures and routes for large shopping malls or theaters (Helbing, 1992; Okazakia & Matsushitaa, 1993; Teknomo, 2002). An application area that is becoming increasingly important is pedestrian navigation. The importance of MMs in navigation is that the more predictable the positioning system states are, the better the measurement outliers can be removed over time. Moreover, if measurements happen to be unavailable for one or more time steps, then the MM allows for prediction of the state estimate.

In the literature, one distinguishes between three different levels of MMs: the mezoscopic, the macroscopic, and the microscopic levels (Helbing, 1992; Teknomo, 2002). A microscopic representation is of more interest in the navigation domain since pedestrian navigation is normally carried out at an individual level. Several pedestrian microscopic MMs exist, such as the Social Force Model (Helbing & Molnar, 1995; Lakoba, Kaup, & Finkelstein, 2005), the Magnetic Force Model (Okazakia & Matsushitaa, 1993), the Benefit Cost Model (Ressel, 2004; Teknomo, 2002), the Cellular Automata Model (Dijkstra, Jessurun, & Timmermans, 2001; Weifeng, Lizhong, & Fan, 2003; Yang, Fang, Li, Huang, & Fan, 2003), and the Queuing Network Model (Osorio & Michel Bierlaire, 2007; Weifeng et al., 2003). The state-of-the-art MMs for navigation applications are random walk models: state-space-based random walks model the pedestrian acceleration using an independent white noise process (Bar-Shalom, Li, & Kirubarajan, 2001); random waypoint models represent the pedestrian's motion by a fixed speed and a randomly selected waypoint (Hyytiä, Lassila, & Virtamo, 2006); Brownian motion uses the model of random movement of particles suspended in a fluid (Banos & Charpentier, 2007); drunkard's walk models use a stochastic model for a drunken person (Sharma & Vishwamittar, 2005); Lévy flight models utilize the step-lengths (or flights) and pause times between flights with uniformly distributed angular directions (Shlesinger, Zaslavskii, & Frisch, 1994); and the stochastic behavioral MM uses the relation between the pedestrian situation and the movement performed (Khider et al., 2012). However, with random walk models, the goal-oriented movement of a pedestrian heading toward a specific destination is not represented. Moreover, experiments have shown that random walk models can take a long time to propose a movement that results in pedestrians entering or leaving rooms, passing through narrow openings, or making sharp turns (Khider, Kaiser, Robertson, & Angermann, 2008). This will result in having some building areas that are very seldom or never explored. The navigation community has widely used a priori knowledge of indoor and outdoor maps to improve the performance of movement models. This has been done, for example, via map matching where position measurements and MMs are related to floor plans and outdoor maps (Brakatsoulas, Pfoser, Wenk, & Salas, 2005). The overall objective is to increase the accuracy of positioning by assuming that pedestrians and vehicles are restricted in movement or position according to streets, paths, and corridors.

Instead of an MM we propose to define a probabilistic map based on the environment, built on the MMs (Kaiser et al., 2011; Khider et al., 2012), to handle multimodality situations. The advantage of defining a probabilistic map with location-dependent probabilistic functions is that it is multifunctional: for instance, it can be used for location-dependent particle weighting in a PF framework to handle multimodality, as a prior map for FootSLAM, or in any MM not only for pedestrians but also, for instance, for other kinds of motions.

### 6.1.5.3 Probabilistic Maps

In this chapter, we describe two ways for generating a probabilistic map. The first method is based on environmental restrictions taking into account floor plans or other forms of maps like outdoor satellite images that provide us with prior information on a location (Section 6.2). The second method is the direct estimation of the probabilistic map during a walk. This can be done via a SLAM algorithm for pedestrians (Sections 6.3 and 6.4).

Knowledge of any prior information about the environment can also help SLAM-based algorithms. This knowledge may be available in two generic forms. First, it may represent prior information on a building layout such as the outer walls or major structures of the building (Kaiser et al., 2012). Second, it might represent prior knowledge of Wi-Fi AP locations or their transmit power or similar features such as the magnetic field. Furthermore, this knowledge can be used to constrain the estimation process of the same kind of feature (e.g., prior building layout used to constrain FootSLAM). But it is also possible to use this prior information to help estimate a different feature (e.g., prior knowledge of Wi-Fi APs or the magnetic field intensity could help the FootSLAM algorithm). Naturally, all combinations of prior information and their usage are possible, at least in principle. One can, for instance, perform WiSLAM with a prior map consisting of a FootSLAM map and a MagSLAM (Robertson et al., 2013) map, which might reduce the computational complexity of WiSLAM, since fewer particles in a PF algorithm might be needed.

## 6.2 Generating a Probabilistic Map from Floor Plans and Satellite Images

In this section, we describe how a probabilistic map based on the environment (e.g., walls, furniture, and outdoor paths) of the pedestrian can be generated. The evaluation of the environment is based on a diffusion algorithm. The motion of a pedestrian is strongly dependent on (1) the physical environment (i.e., walls and obstacles that constrain the motion) and (2) the intention of the pedestrian (i.e., the goal). We concentrate on floor plans and outdoor satellite images to determine a probabilistic map based on the environment. The constraints on the environment restrict the possible heading to only some directions. In this part, we generate a probabilistic map based on the directions a pedestrian might follow, which we call a general probabilistic map. In addition, we describe how a set of goal-oriented probabilistic maps can be obtained.

For indoor pedestrian navigation applications, the probabilistic maps can be used, for instance, for weighting particles in a PF framework, in any MM, or in a SLAM algorithm as a prior map. They are foreseen for applications where prediction of heading is needed, in an indoor-outdoor environment with known floor plans and outdoor maps, and where the possible headings are reduced because of obstacles and walls. After introducing the basic diffusion algorithm, and the extensions to handle outdoor and 3D environments, we describe how to generate a general map and a set of destination-based probabilistic maps.

### 6.2.1 Basic Diffusion Algorithm

The basic diffusion algorithm referred to in Kammann, Angermann, and Lami (2003) is derived from the principle of gas diffusion as studied in thermodynamics and is a standard solution for path finding by robots (Schmidt & Azam, 1993). The idea of the diffusion algorithm is to have a source continuously effusing gas that disperses in free space and that gets absorbed by walls and other obstacles. The core assumption of the algorithm is that a pedestrian will walk along a "sensible" path from the current location to a destination. For a specific destination point in a rectangle area of interest of size $N_x \times N_y$, the diffusion algorithm iterates the following steps:

- The diffusion is expressed by a convolution of the diffusion matrix **D** with the $n \times n$ filter matrix **F** element-wise multiplied by the layout map matrix **L**. Accordingly, the diffusion matrix element at row $i$ column $j$ and at time $k + 1$ is expressed as

$$d_{i,j}(k+1) = l_{i,j} \sum_{p=1}^{n} \sum_{q=1}^{n} d_{i+p-1,j+q-1}(k) \cdot f_{p,q}, \tag{1}$$

where the layout map matrix defines the accessible and inaccessible areas with elements

$$l_{i,j} = \begin{cases} 1 & \text{if } l_{i,j} \text{ is accessible} \\ 0 & \text{if } l_{i,j} \text{ is not accessible} \end{cases} \quad \forall \, i,j : i = 0,\dots,N_x, \; j = 0,\dots,N_y. \tag{2}$$

The coefficients of the filter F of size are defined as

$$f_{p,q} = \frac{1}{n^2} \; \forall \, p,q : p,q = 0,\dots,n, \tag{3}$$

where $p,q$ are the indices of the filter matrix.

- The source is constantly refreshed by forcing $d_{x_d,y_d} \lozenge 1$ at the destination point $(x_d, y_d)$.
- Equation 1 is repeatedly evaluated until the entire matrix is filled with values greater than zero (except for walls and closed areas):

$$d_{i,j} > 0 \; \forall \, i,j : i = 0,\dots, N_x, j = 0,\dots, N_y.$$

The path is computed by backtracking from the destination point toward lower values of the diffusion matrix until the current waypoint is reached. Using a contour plot representation of the diffusion matrix, the path can be calculated following the steepest gradient (i.e., crossing the contour lines perpendicularly) until the destination is reached.

Details on the basic diffusion algorithm can be found in Schmidt and Azam (1993) and Kammann et al. (2003). A color-coded plot representing the resulting gas diffusion in space for a destination point at the center of the corridor of the second floor of our new office environment is shown in Figure 6.2.

## 6.2.2 Extending to Outdoor Environments

For outdoor areas public information rarely exists describing walkable areas, fields, parks, flower beds, and so on. Street maps may help people obtain information about road layout, but the areas in front of buildings are not covered. Outdoor maps contain useful information on constraints influencing pedestrian movement such as areas with different degrees of accessibility and "sensible" outdoor paths. Such accessibility information can be used to extend the MM to propose reasonable movements in outdoor areas. Outdoor

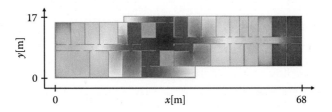

**FIGURE 6.2**
A color map representing the resulting diffusion matrix for a source (destination point) at the center of the corridor of the second floor of an office environment. Red represents higher diffusion matrix values. Walls and inaccessible areas are depicted in gray.

maps can be obtained via any map provider in different formats such as satellite images, street maps, and topographical maps. The layout map matrix **L** that was used in the computation of the diffusion matrix in the previous section has been modified to handle the different degrees of accessibility provided by outdoor maps. The elements of the new layout map matrix **L** are defined as (Kaiser et al., 2012)

$$l_{i,j} = \begin{cases} \dfrac{1}{\upsilon} & \text{if } l_{i,j} \text{ is accessible,} \upsilon = 1\ldots255 \\ 0 & \text{if } l_{i,j} \text{ is not accessible} \end{cases} \quad \forall\, i,j : i = 0,\ldots N_x, j = 0,\ldots,N_y, \quad (4)$$

where $\upsilon$ represents the accessibility level. The most accessible areas will have values of $\upsilon = 1$, whereas the least accessible areas will have values of $\upsilon = 255$. For other points, the layout map matrix will have values between 1 and 1/255 depending on the accessibility level. The 255 accessibility levels proved sufficient to represent the different accessibility levels of various areas.

To obtain a layout for the outdoor area, we proposed that the layout information be extracted from satellite photos of the area under consideration. The layout information can easily be obtained via satellite images. Let us consider the map example depicted in Figure 6.3a. Accessibility information in Figure 6.3a can be obtained by analyzing the different colors present. For instance, forest areas are usually dark green and meadows light green, while streets and paths are usually light gray. The type of area that any point in the image belongs to is identified by its respective color. More sophisticated image segmentation algorithms like partial-differential-equation-based methods, histogram-based methods, or edge-detection algorithms can also be considered to detect different types of areas from satellite images. Figure 6.3b (from Kaiser et al., 2012; www.tandfonline.com) shows the results of the resulting layout with marked shadows. The walkable area is given in white, the less accessible area in dark gray, the shadowed area in light gray, and the walls in black. Two inaccessible office areas exist, as can be seen in the figure.

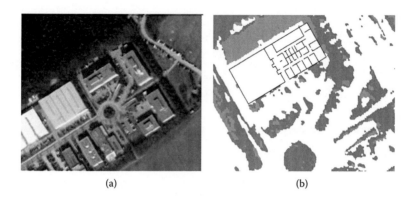

(a)                                                                              (b)

**FIGURE 6.3**
a. Satellite photo of an office environment (from "DLR/Digital Globe Provided by European Space Imaging"). b. Layout (from Kaiser et al., 2012; www.tandfonline.com) generated from a cutout of the satellite photo. Walkable area is given in white, less accessible area in dark gray, shadowed area in light gray, and the walls in black.

## 6.2.3 Extending to Multifloor Environments

Many indoor environments have more than one story or floor. In such cases, it is necessary to consider several floor plans and, accordingly, different heights. One option is to allow the gas to flow in three dimensions, which means that the diffusion algorithm is also extended to 3D. However, this would be unrealistic for an MM since pedestrian motion does not follow the third dimension freely. Pedestrians walk on the ground within each story. Thus, a multifloor building is treated as a combination of individual 2D floors. However, difficulties arise in the stair area, because of the change in the ground height at each step. To simplify the model, in our case we projected the stairs into the 2D area, where the stairs are connected to two different levels. Finally, our building consists of rectangular areas representing different floor levels and additional rectangular areas for each projected stair.

When we apply the gas diffusion process, the gas is allowed to flow between floor levels only through stair areas. To consult the effect of the stair areas on the gas distribution, during calculation of the diffusion matrix for a specific floor level, we included the respective stair areas in that level before the diffusion process was applied. Since the floor level might be connected to two floor levels, as is usually the case in staircases, the stair areas might overlap. Therefore, the diffusion calculation was adapted to the overlapping areas by applying the filtering process of each level twice, including the respective stair area. With this, we guarantee that the gas flows in each level, and we can reduce complexity by applying this quasi-3D diffusion algorithm. More details on the 3D extensions can be found in Khider, Kaiser, Robertson, and Angermann (2009). The methodology followed to calculate the altitude at each time step was to generate a matrix containing a relevant

$z$ value (altitude) for every $(x, y)$ position. The projection of the stairs in 2D allows the altitude information for the stair areas to be appended. So, for each step area of the staircase a different altitude can be stored.

### 6.2.4 Generating the Probabilistic Map

For obtaining a probabilistic map based on the a priori knowledge of floor plans, we assume that the source of the gas is the *current waypoint,* and we calculate *angular probability density functions (PDFs)* out of the gas distribution. No path-finding algorithm is needed, and the weighting is independent of the choice of destination points. A probabilistic map resulting from such an approach is referred to as a *general probabilistic map.*

In case of the availability of a priori knowledge of possible destination in addition to the floor plan, the probabilistic map is generated by using the diffusion algorithm in its original fashion with the central assumption to have a source-effusing gas, which is one of possible destination points similar to those in the destination-based MM (Khider et al., 2009). Here, a path finder is needed to find the path to that destination point. In this case, a set of destination-based probabilistic maps will be obtained (i.e., a probabilistic map for each destination point).

#### 6.2.4.1 A General Probabilistic Map

For generating the general probabilistic map, we determine the so-called angular PDFs (Kaiser et al., 2011) for each waypoint. One advantage of taking the current waypoint as the source of the gas rather than the destination point is that it is then possible to calculate a weighting function for the directions that can be followed directly from the gas distribution. Another advantage is that the path-finding algorithm is no longer needed, and the weighting is independent of the choice of destination point location. In addition, the diffusion rectangular area can be restricted to a small area around the current waypoint, much reducing the computational effort. With such a low computational effort, one can calculate the PDF values for new regions during the runtime of real-time implementations.

The diffusion matrix calculation is derived from the diffusion algorithm described in Section 6.2.1. We use a sliding square window of size $N_x \times N_x$ ($N_x$ is odd-numbered), with the current waypoint $(x_m, y_m)$ at its center. Depending on the grid size and the size of the square window, the resolution of the angular PDF can be varied. If the current waypoint is the source of the gas, angular PDFs are defined, assuming that the possible headings follow the gas distribution. For each waypoint the diffusion matrix contains the gas concentration values $\mathbf{D}_m$ is precomputed. Within the diffusion matrix, a wall-bounded contour line can be obtained by selecting all diffusion values that are below a certain threshold $T$. The diffusion values on this line are equal along the line except at points where walls stop the gas flow. A wall-bounded contour line

**Polar Chart**

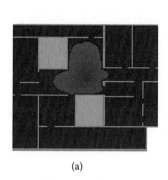 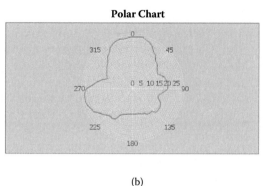

(a) (b)

**FIGURE 6.4**
a. Wall-bounded contour line (dark red) of the diffusion values of Figure 6.2 marked in dark red with threshold = 0.001. Only a cutout of the building is shown in the figure. The source of the gas is in the center of the corridor at the same location as in Figure 6.2. b. Resulting polar chart for the angular PDF for the same location as in Figure 6.4a.

generated from the diffusion values of Figure 6.2 with a threshold $T = 0.001$ is shown in Figure 6.4. From this wall-bounded contour line, we obtain the angular PDF using the Euclidean distances from the actual waypoint ($x_m$, $y_m$) to the points of the wall-bounded contour line. These points are of different distances to the actual waypoint depending on the considered angle ρ. Normalizing these distances results in a PDF of the different angles that can be followed by the pedestrian from the waypoint. The complete derivation of the angular PDFs can be found in Kaiser et al. (2011).

If we store the angular PDFs of all waypoints in a map, we obtain a general probabilistic map for possible directions at each discrete location ($x,y$). The probabilistic map gives us information about possible directions at each waypoint. Precisely, for each possible waypoint the respective angular PDF is stored at the location of that waypoint. The resolution of the PDFs is restricted to the resolution of the map and the resolution of the calculated angles.

In our experiments we used a 5° spacing for angle ρ, $0 \leq \rho < 360°$, generating 72 discrete values. This manifested to be sufficient to obtain a smooth weighting function while minimizing the volume of data to be stored. Our probabilistic angular PDF map **M** of a rectangle area of size $N_x \times N_y$ comprises the random variable vectors

$$\vec{M}_{x,y} = \left\{ M_{x,y}^1, \ldots, M_{x,y}^{N_\rho} \right\} \quad \forall \quad x = 1, \ldots, N_x, \quad y = 1, \ldots, N_y, \tag{5}$$

where $N_\rho$ is the number of discrete values of the angular PDF to be stored depending on the spacing for angle ρ. If we apply the angular PDFs to a FootSLAM map that consists of hexagons, we reduce $N_\rho$ to 6 because of the six edges of the hexagons. Supplementary details of calculating the angular PDFs and mapping them onto the FootSLAM map are given in Kaiser et al. (2012).

We can use the angular PDFs in either a normalized form or an unnormalized form. The normalized fashion is suitable for weighting particles or for an MM. If we use the angular PDF in an unnormalized fashion, the angular PDFs values are presented by the absolute distances to the contour lines of the diffusion matrix. The probability of visiting a location is inherently given by the integral overall angular PDF values of all possible directions of that waypoint in relation to the integral overall angular PDF values. This information is especially useful when applying the angular PDFs as a prior map for FootSLAM.

The probability of visiting a location $(x, y)$ can be obtained as follows:

$$\pi_{x,y} = \frac{\sum_{i=1}^{N_\rho} \hat{\mathbf{M}}_{x,y}^i}{\sum_{x=1}^{N_x} \sum_{y=1}^{N_y} \sum_{i=1}^{N_\rho} \hat{\mathbf{M}}_{x,y}^i} \tag{6}$$

where $\hat{\mathbf{M}}_{x,y}$ is our unnormalized angular PDF function at location $(x, y)$ that is a function of angle $\rho$. Because of discretization, the integral is substituted by the sum.

### 6.2.4.2 Destination-Based Probabilistic Maps

The probabilistic map can be built based on the destination-based MM. The core assumption of the destination-based MM is that a pedestrian will walk along a "sensible" path from the current location to a random destination with a given probability to change this destination before reaching it. Therefore, we specify a set of $N_d$ destination points for the two-dimensional rectangular area in which the pedestrian walks. For each destination point $m$ we precompute the diffusion matrix $\mathbf{D}_m$ and an angle matrix $A_m$ in order to keep the computation effort low during the run. The angle matrix stores the angles of the direction of the paths from all possible locations to that destination point.

In free space, the path-finding approach, based on the work by Kammann et al. (2003), proposes paths that are not straight lines from the current waypoint to the destination. However, without obstacles, pedestrians follow straight lines to a destination in free space. Therefore, the tracing approach was modified in order to obtain more realistic paths. Details of the modified tracing approach can be found in Khider et al. (2012).

If we use the destination-based diffusion MM, for each destination we obtain a probabilistic map for the possible directions heading to that destination. Thus, the result is a set of probabilistic maps $\bar{\mathbf{M}} = \{\mathbf{M}^1, \ldots, \mathbf{M}^{N_d}\}$. Each probabilistic map contains at each location $(x, y)$ a Gaussian function having a mean equal to the direction proposed by the destination-based MM. The variance of the Gaussian function is set based on the number of walls in the vicinity; more walls lead to lower variance. This is based on the fact that it is more probable that a pedestrian follows a specific path if more walls are in the vicinity.

### 6.2.5 Incorporating Probabilistic Maps in PF-Based Position Estimators

In PF-based position estimators, where the prior transition is used for the proposal function and the likelihoods are based on the measurements, the general probabilistic map can be used as an importance function to propose a realistic pedestrian direction depending on the local environment independently from destination points. Normalized angular PDFs can be used for this purpose. Moreover, the destination-based probabilistic maps can be used as an importance function to compute realistic pedestrian paths between the location of a particle and the destination point that is part of the state space of the particles. In both cases, the speed can be modeled by a random process, for example, the stochastic behavioral MM (Khider et al., 2012).

In PF position estimators proposing from measurements and weighting with the MM, the resulting PDFs of the general probabilistic map or the destination-based probabilistic maps can be used directly to weight particles at every time step. As pedestrians are not always moving according to the surrounding constraints, one might want to model additional stochastic pedestrian behaviors. In such cases, an alternate model that switches between motion based on one of the predeveloped models and a random walk can be used. Khider et al. (2009) used a Markov model for this purpose.

## 6.3 A Probabilistic Map of Human Motion Based on FootSLAM

FootSLAM (Robertson et al., 2009) was proposed as a means of generating a map based on noisy odometry measurements taken from a single or multiple pedestrians. To date, FootSLAM implementations have been based on a first-order Markov process, meaning that the probability distribution of future motion according to the model is conditioned only on the present location but not on the past. FootSLAM builds on the algorithmic principle of SLAM (Durrant-Whyte & Bailey, 2006) since the location of the pedestrian is unknown and is jointly estimated with the probabilistic MM. In FootSLAM, the term *map* is used to denote this model. It encodes the observations of a person's next motion step at each location in the environment.

### 6.3.1 Mapping and Localization

FootSLAM is rooted in a Bayesian recursive formulation (Robertson et al., 2009) whereby at each time step $k$ the pedestrian takes a step $U_k$ based on his or her intent $Int_k$ (directed by his or her visual cues $Vis_k$) moving from pose (his or her position in 3D and heading) $P_k$ to $P_{k+1}$. We rely on a foot-mounted IMU to collect measurements of a pedestrian's steps $Z_k^U$, having no direct access to the intent or visual cues of the pedestrian. The IMU measurements are preprocessed by the unscented Kalman filter (UKF) (Zampella et al., 2012) to generate what we call *human* odometry (measurements of a pedestrian's steps). These odometry data are still subject to correlated errors $E_k$ being the most important source of error in the heading drift (Robertson et al., 2009).

The goal of FootSLAM is the estimation of the full posterior (Robertson et al., 2009):

$$p\left(\left\{\boldsymbol{PUE}\right\}_{0:k}, \boldsymbol{M} \mid \boldsymbol{Z}_{1:k}^U\right).$$

To do this, we employ a PF (Arulampalam, Maskell, Gordon, & Clapp, 2002) to track different hypotheses for the odometry errors, and we discretize the space into a grid of uniform and adjacent hexagons. The grid is composed of $N_x$ rows and $N_y$ columns of hexagons (Garcia Puyol et al., 2013). Each hexagon has six edges $e \in \{0...5\}$ and a radius $r$. Following the FastSLAM factorization (Montemerlo, Thrun, Koller, & Wegbreit, 2002), for each particle $i$ we store its own map. A particle's map consists of the hexagons visited by the particle along with its *transition counters*; that is, the number of times that it has crossed each one of the edges of the hexagon $C_h^e$:

$$M^{[i]} = \left\{h, \boldsymbol{C}_h\right\},$$

with $\boldsymbol{C}_h = (\mathbf{C}_h^0, \mathbf{C}_h^1, \mathbf{C}_h^2, \mathbf{C}_h^3, \mathbf{C}_h^4, \mathbf{C}_h^5)$.

This is based on the assumption that human motion follows a random walk of order one, that is, the current position of the pedestrian depends solely on his or her previous position. The two main steps of the PF algorithm are the proposal step and the weighting step. In the proposal step, each particle proposes from a suitable error distribution (Robertson et al., 2009). Particles move according to the measured step $Z_k^U$ plus the proposed errors, visiting and crossing the edges of different hexagons. In the weighting step, particles are weighted based on the transition counts of the edge $e$ of the hexagon $h$ they have crossed when moving from $P_{k-1}$ to $P_k$:

$$w_k^{[i]} = w_{k-1}^{[i]} \left\{ \frac{C_h^e + \alpha_h^e}{C_h + \alpha_h} \right\}^{[i]}, \tag{7}$$

where

$$C_h = \sum_{e=0}^{e=5} C_h^e \text{ and } \alpha_h = \sum_{e=0}^{e=5} \alpha_h^e$$

represent any prior available information regarding the transition probabilities. The number 1 means that particles that revisit transitions will obtain a greater reward than particles that do not. Thus, when a pedestrian closes a loop, particles that have estimated the correct drift will obtain a greater weight over time, and particles with a low weight will be deleted during the resampling step (Robertson et al., 2009). As a result, a map of the accessible areas of the environment is generated. Garcia Puyol et al. (2013) introduced a tree-based data structure called *H-tree* that allows for efficient map updates, ensuring $O(t \log t)$ complexity growth for FootSLAM with a growing area, where $t$ is the time. Garcia Puyol, Bobkov, Robertson, and Jost (2014) extended FootSLAM to multistory environments using a grid of adjacent hexagonal prisms.

## 6.3.2 Collaborative Mapping

As stated before, the pedestrian must revisit areas for a suitable map to emerge. To avoid making this requirement, we can resort to *crowdsourced* mapping. In crowdsourcing (Howe, 2006) we delegate the mapping task to a group of collaborating pedestrians. The advantage of such an approach is that we can use the visits of a pedestrian as prior knowledge for the FootSLAM estimation process of another pedestrian, generating a more accurate map. In addition, the map extent will be more complete since different pedestrians visit different areas. Our envisioned approach is called FeetSLAM (Robertson et al., 2011) and is based on "Turbo" codes: at each iteration and for each collected odometry data set, the combined map of all other data sets is computed and used as a prior map in the next iteration for that given data set. Given the rotation- and translation-invariant property of SLAM algorithms, to merge the individual maps into one single combined map, we need to find the geometric transformation (angular and spatial displacements) that places all of them within the same coordinate system (Robertson et al., 2011).

Figure 6.5 shows an example of a combined map obtained from four data sets collected in an office environment. Garcia Puyol et al. (2013) showed that this algorithm is suitable to generate an indoor map database of the whole indoor world within a couple of years even with a low proportion of collaborating mapping individuals and that this can be achieved with low memory requirements. This database could in turn be used by pedestrians with localization needs, for example, multimodal travelers (Garcia Puyol et al., 2014) or for emergency and security applications (Garcia Puyol, Frassl, & Robertson, 2012). Future work will address crowdsourcing the generation of radio maps (Section 6.4) and magnetic maps.

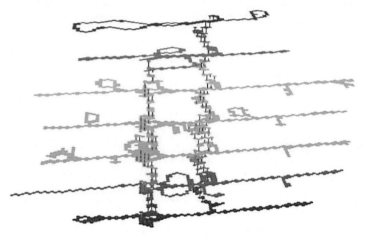

**FIGURE 6.5**

Combined map (on the bottom) obtained from four data sets collected in an office building (Building 103 in Oberpfaffenhofen, Germany, on the top). The pedestrians visited all office floors, the basement, and the terrace. Different colors encode different floors. Vertical transitions are shown as rectangular polyhedrons and mark the location of the two sets of stairways.

## 6.4 A Probabilistic Map Based on WiSLAM

The RSS measured from transmitters of a wireless local area network (WLAN), for example, Wi-Fi APs, is a useful source of information for indoor positioning. Although the network design is usually driven by transmission issues; that is, each point of the indoor area needs to be included in only one AP's coverage area, redundancy is very common in practice, because, for instance, of the simultaneous presence of different networks, and therefore, multiple AP coverage is usually obtained. The WLAN map, in order to be used by positioning algorithms, consists either of the extensive radio map of the environment, which is built based on supervised RSSs measured at a dense set of locations, or of a compact representation of the radio environment, for instance, position and transmit parameters of all APs, as well as some propagation features. Bruno and Robertson proposed a Bayesian SLAM algorithm based on RSS, called WiSLAM, which exploits the compact representation of the environment (Bruno & Robertson, 2011, 2013). It is based on the FastSLAM factorization and fully integrated with FootSLAM, since each RSS yields a further factor in the particle weights.

In WiSLAM, radio propagation is assumed to follow the well-known pathloss model, whereby the expected signal strength $h(r)$ in dBm received at distance $r$ from an AP is (Parsons, 2000)

$$h(r) = h_0 - 20\alpha \log\left(r/r_0\right),$$

where the logarithm is decimal, $r_0 \sim 1$ m is a fixed distance, $h_0 = h(r_0)$ will be referred to as the "transmit power," and $\alpha$ is the decay exponent. The actual RSS is a noisy measurement of $h(r)$, and its randomness is usually described by a Gaussian, Rice, or Rayleigh distribution. In addition, independence of noise across both time and APs is often assumed.

Given the user's position, transmit power, and decay exponent, each measurement provides information on the AP's position. It is represented as a circular PDF, centered at the user's position, with radius $r$ and with uncertainty around $r$ related to the variance of the noise. When $k$ measurements are available, the AP's position PDF is given by the normalized product of all circular PDFs, each one centered in a different position, reflecting the user's changing location. Note that such a PDF is multimodal for small $k$. For instance, when the user walks a straight line, the resulting PDF is symmetrical with respect to the trajectory. Furthermore, the width of the AP's PDF depends on $k$ and on the proximity of the AP to the user's trajectory; that is, the algorithm extracts as much information as available in the measurements and no more, avoiding dangerous overconfidence issues.

To achieve computational tractability, we approximated the AP's position PDF in WiSLAM by a Gaussian mixture model, with a suitable number of components, which is trained at some fixed $k$ numerically. Bruno and Robertson (2011) showed that the Gaussian mixture model approximation allows an update step of the PDF at each new measurement that presents a time-linear complexity. Ideally, the transmit power and the decay exponent should be known in advance. When this is not the case, they can be introduced in the map as additional parameters to be estimated. Bruno and Robertson (2013) discretized both parameters on a grid, representing a set of hypotheses, each consisting of a pair of values per AP. An independent PDF of the AP's position is computed as before under each hypothesis and is

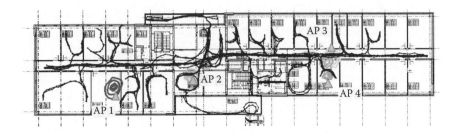

**FIGURE 6.6**
Example of a map obtained with WiSLAM. The estimated trajectory is in blue, and the contour lines show the AP's position PDFs. The ground truth is represented by the floor layout, and the AP's true positions are denoted by green triangles.

weighted with the hypothesis probability. With each new RSS measurement, an update of the hypothesis probabilities and all the conditional AP position PDFs is required.

A map obtained by WiSLAM is presented in Figure 6.6, based on a data set collected during a roughly 30-minute walk in a $65 \times 35$ m$^2$ floor of a building. The sensors were a foot-mounted IMU for inertial measurements and a handheld smartphone for RSS. The layout of the floor and the exact position of the APs, denoted by green triangles, are depicted as ground truth, but not employed at all in the algorithm. The AP's position PDFs, estimated at the end of the walk, are depicted by contour lines at 1%, 10%, 50%, 75%, and 99% of the maximal value of each AP's PDF. Remarkable accuracy is shown in estimating both the trajectory and the AP's PDF. Other examples are given in Bruno and Robertson (2011, 2013).

## 6.5 Quantifying Maps and Experimental Results

### 6.5.1 Experimental Settings

In the experiments, we applied FootSLAM on the second floor of our office environment (see Figure 6.2). A pedestrian walked 25 minutes wearing a low-cost foot-mounted IMU (XSENS MTx). After preprocessing the data by a step detector based on ZUPTs and a Kalman filter (Zampella et al., 2012), the data were used as input for the offline-executed FootSLAM algorithm. The number of particles used in the RBPF is $N_p = 10000$ and the radius of the hexagons is $r = 0.5$ m. The floor-plan-based probabilistic map is the general probabilistic map based on the angular PDFs with a 5° spacing for discretized angle $\rho$, and the general map is transformed to a FootSLAM map with a 60° spacing (Kaiser et al., 2012).

## 6.5.2 Entropy Comparison

To compare probabilistic maps, Kaiser, Garcia Puyol, and Robertson (2013) derived two different entropy metrics using information-theoretic measures. The two metrics are the *map entropy* and the *step entropy rate* conditioned on the history of poses. The map entropy represents the uncertainty of the map itself, whereas the step entropy rate is the rate of growth of the step entropy; that is, the uncertainty regarding each next step. These uncertainty metrics are useful for map comparison, map combination, map selection, and theoretical analysis. This is also needed in collaborative mapping (FeetSLAM, see Section 6.3.2). In this section, we compare the floor-plan-based probabilistic map with the FootSLAM-estimated probabilistic map.

Kaiser et al. (2013) showed the derivation of both the map entropy and the step entropy rate. Both metrics are similar; only the range of values is larger with the differential map entropy and it considers also the number of times a hexagon was visited. In this section, we show only the step entropy rate and refer to Kaiser et al. (2013) for the map entropy. The derivation of the step entropy rate is based on the assumption that the sequence of pedestrian poses can be regarded as a Markov chain with stationary distribution $\pi$. The stationary distribution represents the probability of hexagon visits. With the *random walk on a weighted graph* theory, it can be shown that the stationary distribution is

$$\pi_h = \frac{C_h + \alpha_h}{C_M + \alpha_M}$$

where $\mathcal{H} = \{h_0, h_1, \ldots, h_{N_{h-1}}\}$ is the set of $N_h$ hexagons of map $M$,

$$C_M = \sum_{i \in \mathcal{H}} C_i \text{ and } \alpha_M = \sum_{i \in \mathcal{H}} \alpha_i.$$

It can be shown that the step entropy rate $H(\mathcal{P})$ of the Markov chain $\mathcal{P}$ can be derived as (Kaiser et al., 2013)

$$H(\mathcal{P}) = \lim_{k \to \infty} H\left(P_k \mid P_{k-1}\right) = \sum_{h \in \mathcal{H}} \pi_h \sum_{e=0}^{5} H\left(P_k \mid P_{k-1}\right)$$

$$= -\sum_{h \in \mathcal{H}} \frac{C_h + \alpha_h}{C_M + \alpha_M} \sum_{e=0}^{5} \frac{C_h^e + \alpha_h^e}{C_h + \alpha_h} \log \frac{C_h^e + \alpha_h^e}{C_h + \alpha_h}$$

(8)

Figure 6.7 shows the estimated FootSLAM map from a relatively long walk (25 minutes). Naturally, the FootSLAM walk does not cover the whole area of the second floor of our building, but the whole area is covered with the

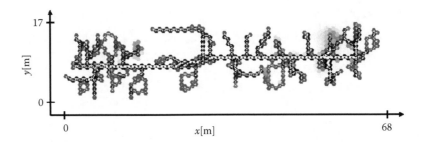

**FIGURE 6.7**
FootSLAM estimated map of the second floor of an office environment (see Figure 6.2). This is
the result of a 25-minute walk through the second floor of the building.

map resulting from the floor plan interpretation (at each location of the second floor of Figure 6.2, an angular PDF like that of Figure 6.4 is available). On the other hand, the floor plan interpretation does not take into account obstacles like tables and cupboards. The information about obstacles could be included in the floor plan via the layout matrix, but the effort seems to be high, and furniture can be moved. The values for the step entropy rate are shown in Table 6.1. The comparison is done for the same number of hexagons covering the whole second floor of the building. Because all hexagons are covered by the general probabilistic map, the entropy is better (lower) than it is for the estimated map, where several not-covered hexagons are counted with maximum entropy. The entropy (2.48 bits) is maximal when the six-edge transitions are equally probable.

To illuminate another difference between the two kinds of probabilistic maps, namely, the floor-plan-based map and the FootSLAM-estimated probabilistic map, and to compare them at a different aspect, we additionally investigated how often an area is visited based on the respective probabilistic map. Figure 6.8 shows the resulting probabilities for the angular PDF map of our office building. From this figure we can see that the walks concentrate on big rooms and not in corridors as it would be expected in a typical office building. The cause for this is that the probability of a walk leaving the room is small, because the angular PDFs have not very high values in the direction of the doors, and inside the room the angular PDFs are equally distributed.

**TABLE 6.1**

Step Entropy Rate Values for the General Probabilistic Map Obtained with the Diffusion Algorithm and the FootSLAM-Estimated Map

|  | **Step Entropy Rate** |
| --- | --- |
| General probabilistic map | 2.37 bits |
| FootSLAM-estimated map | 2.44 bits |

*Note:* The values are computed over the same area (i.e., the same number of hexagons).

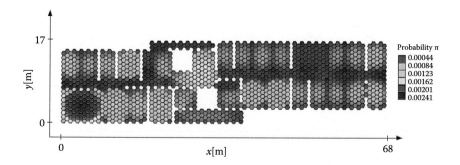

**FIGURE 6.8**
Probability of hexagon visits of the general probabilistic map based on angular PDFs transformed to a hexagon raster.

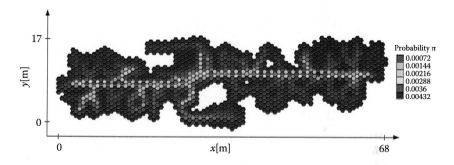

**FIGURE 6.9**
Probability of hexagon visits of the FootSLAM map based on real pedestrian walking (4 hours total walking time), where the angular PDF map was used as a prior map with known starting conditions.

This figure is based on a hexagon raster because we transformed the angular PDF map to a FootSLAM map.

For comparison, Figure 6.9 shows the resulting probability distribution derived from real pedestrian walking (1 hour total walking time) transformed the FeetSLAM aggregated posterior map to fit the walks into the building area. One can see that the walk concentrates on the corridors and some of the rooms the person visited. This real walk is more like the real behavior of (permanently) walking humans: walks through corridors and entering and leaving some rooms (staying in rooms is not considered for these evaluations). This behavior is not reflected in the probability distribution of the angular PDF probabilistic map.

### 6.5.3 Positioning Error Comparison

Finally, we compare the positioning accuracy of FootSLAM with and without the use of our general probabilistic map based on angular PDFs. Since

FootSLAM maps are obtained allowing the particles to explore different hypotheses, they are rotation invariant. Usually the starting conditions are assumed not to be perfectly known, so the resulting FootSLAM map may vary in orientation and final position. To compare the FootSLAM maps with and without the use of a prior map, we used FootSLAM for this purpose with *perfectly known starting conditions.* The angular-PDF-based probabilistic prior map is used with a relatively strong prior strengthening factor (the transition counters are multiplied with the prior strengthening factor in order to increase/decrease the influence of the prior map). In Figure 6.10a and 6.10b, we depicted the FootSLAM results for two different walks: an 8-minute walk (two same loops within the second floor of our building) and a 4-minute walk (only one loop, same area), respectively. The error is computed against predefined ground truth points (GTPs) that were measured to subcentimeter accuracy with a Leica Disto D210. Because FootSLAM maps are rotation invariant, especially at the beginning of the walk, it may happen that the best hypothesis (errors are measured during the run and not from the final best map) is rotated around the starting point. Figure 6.10a shows the results for the long walk, where the map was converged to the correct rotation after 250$s$ (second loop within the building). Therefore, the position error decreases after this time point. For comparison, we depicted also the error values for the transformed estimated trajectory. Here we transformed the positions at the GTPs to minimize the average total error. Because the error was only high at the beginning, after transformation, the error increases at the end. The results for the use of our angular PDFs as prior map show that we converge very quickly (less positioning errors at the beginning). With this, we are able to obtain a positioning error of 0.6$m$.

Figure 6.10b shows the results for the short walk, where the rotation of the map is not corrected during the walk when using no prior map. The mean error of this relative short walk is 1.31$m$. If we transform the map to fit best to the GTPs, we obtain much better results (0.36$m$) because the whole estimated trajectory was rotated. On the other hand, if we use the prior map, we have very good results from the beginning without the need for transformation (0.62$m$). It should be noted that the values for the transformation are usually not known and are used here only to illustrate the different estimation accuracies achieved.

## 6.6 Conclusions

In this chapter, we introduced map-aided indoor navigation. Instead of using maps and floor plans only as a simple binary constraint, we discuss the use of *probabilistic maps* as an element of sequential estimation algorithms. We distinguished between probabilistic maps derived from existing floor plans

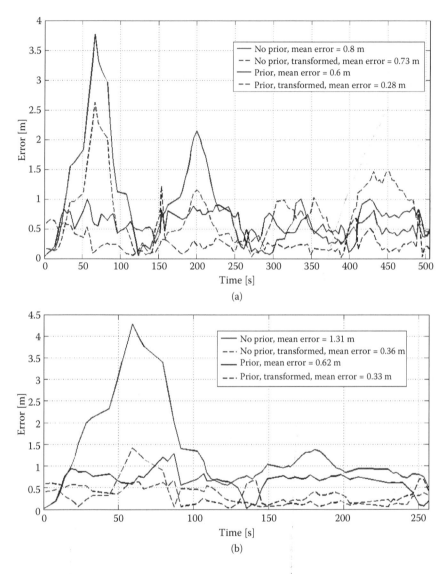

**FIGURE 6.10**
Positioning error over time for FootSLAM with and without the use of a floor-plan-based prior map. The error is evaluated at given ground truth points. Figure 6.10a shows the results for a long 8-minute walk (i.e., two loops within the second floor of our office environment). The dashed curves are obtained after transformation of the trajectory to minimize the average error. Figure 6.10b shows the results for a shorter walk (4 minutes, one loop).

and probabilistic maps that are estimated from sensor data collected from people walking around a building. The floor-plan-derived maps can be obtained using a diffusion algorithm, where an angular PDF is computed for each possible location. This is an estimate of the probability distribution of a pedestrian's next step direction at that location. From collected sensor data, the probabilistic maps can be estimated using a SLAM algorithm such as FootSLAM. In the absence of other sensors or signals and relying only on low-cost inertial sensors or other forms of PDR, we believe that knowledge of such maps is an essential element to achieve high location accuracies. We showed how simplistic binary constraints derived from floor plans can fail under some circumstances, especially when multimodal posteriors occur. We presented the advantage of using preexisting maps even when performing SLAM, where faster convergence and higher accuracies are obtained. We briefly summarized WiSLAM, which represents a map of radio APs in an environment. While WiSLAM required additional sensors and the presence of radio infrastructure, both the ubiquity of mobile phones with WLAN receivers and the high density of APs in buildings make a WLAN map a very useful additional and feasible feature of an indoor map.

## Acknowledgment

We would like to thank Michael Angermann for his encouragement and support.

## References

Anon. (2006). *Ekahau positioning engine*. Retrieved from http://www.ekahau.com.

Arulampalam, S., Maskell, S., Gordon, N., & Clapp, T. (2002). A tutorial on particle filters for on-line non-linear/non-Gaussian Bayesian tracking. *IEEE Transactions on Signal Processing, 50*(2), 174–188.

Bahl, P., & Padmanabhan, V. N. (2000). Radar: An in-building RF-based user location and tracking system. In *INFOCOM 2000: Nineteenth annual joint conference of the IEEE computer and communications societies*. New York, NY: IEEE.

Banos, A., & Charpentier, A. (2007). Simulating pedestrian behavior in subway stations with agents. In *Proceedings of the 4th European Social Simulation Association, 2007* (pp. 611–621). Toulouse, France.

Bar-Shalom, Y., Li, X. R., & Kirubarajan, T. (2001). *Estimation with applications to tracking and navigation*. Hoboken, NJ: John Wiley & Sons.

Beauregard, S., Widyawan, & Klepal, M. (2008). Indoor PDR performance enhancement using minimal map information and particle filters. In *Proceedings of the IEEE/ION position, location and navigation symposium, 2008 IEEE/ION*. New York, NY: IEEE.

Brakatsoulas, S., Pfoser, D., Wenk, C., & Salas, R. (2005). On map-matching vehicle tracking data. In *VLDB 2005 Proceedings of the 31st international conference on very large data bases* (pp. 853–864). VLDB Endowment.

Bruno, L., & Robertson, P. (2011). WiSLAM: Improving FootSLAM with WiFi. In *Proceedings of the International IEEE conference on indoor positioning and indoor navigation IPIN'11*. New York, NY: IEEE.

Bruno, L., & Robertson, P. (2013). Observability of path loss parameters in WLAN-based simultaneous localization and mapping indoor positioning. In *Proceedings of the IEEE international conference on indoor positioning IPIN 2013*. New York, NY: IEEE.

Chiu, D. S., & O'Keefe, K. P. (2008). Seamless outdoor-to-indoor pedestrian navigation using GPS and UWB. In *Proceedings of the 21st international technical meeting of the satellite division of the Institute of Navigation*. Manassas, VA: Institute of Navigation.

COST Action 231. (1999). *Digital mobile radio towards future generation systems: Final report*. European Commission.

Dijkstra, J., Jessurun, A. J., & Timmermans, H. J. (2001). *A multi-agent cellular, automata model of pedestrian movement, pedestrian and evacuation dynamics*. Berlin, Germany: Springer-Verlag.

Durrant-Whyte, H., & Bailey, T. (2006). Simultaneous localization and mapping: part I. *Robotics and Automation Magazine, IEEE, 13*(2), 99–110.

Ferris, B., Fox, D., & Lawrence, N. (2007). WiFi-SLAM using Gaussian process latent variable models. In *Proceedings of the International Joint Conferences on Artificial Intelligence IJCAI, 2007* (pp. 2480–2485). Menlo Park, CA: IJCAI.

Foxlin, E. (2005). Pedestrian tracking with shoe-mounted inertial sensors. In *IEEE computer graphics and applications*. New York, NY: IEEE.

Frank, K., Krach, B., Catterall, N., & Robertson, P. (2009). Development and evaluation of a combined WLAN and inertial indoor pedestrian positioning system. In *Proceedings of the 22nd international technical meeting of the satellite division of the Institute of Navigation*. Manassas, VA: Institute of Navigation.

Frassl, M., Angermann, M., Lichtenstern, M., Robertson, P., Julian, B., & Doniec, M. (2013). Magnetic maps of indoor environments for precise localization of legged and non-legged locomotion. In *Proceedings of the IEEE/RSJ International conference on Intelligent robots and systems (IROS), 2013*. New York, NY: IEEE.

Galileo Ready Advanced Mass Market Receiver. (2009). *Market definition and court technology report* (Technical report, GSA-227890 D1.1 version 1.6). EU Project.

Garcia Puyol, M., Bobkov, D., Robertson, P., & Jost, T. (2014). Pedestrian simultaneous localization and mapping in multistory buildings using inertial sensors. *Transactions on Intelligent Transportation Systems, 15*(4), 1714–1727.

Garcia Puyol, M., Frassl, M., & Robertson, P. (2012, September). Collaborative mapping for pedestrian navigation in security applications. *Future Security*, 49–60.

Garcia Puyol, M., Robertson, P., & Angermann, M. (2013). Managing large-scale mapping and localization for pedestrians using inertial sensors. In *Proceedings of the IEEE international conference on pervasive computing and communications workshops (PERCOM workshops)* (pp. 121–126). New York, NY: IEEE.

Garcia Puyol, M., Robertson, P., & Heirich, O. (2013). Complexity-reduced FootSLAM for indoor pedestrian navigation using a geographic tree-based data structure. *Journal of Location Based Services, 7*(3), 182–208.

Hardegger, M., Roggen, D., Mazilu, S., & Troster, G. (2012). ActionSLAM: Using loca-tion-related actions as landmarks in pedestrian SLAM. In *Proceedings of the IEEE international conference on Indoor positioning and indoor navigation IPIN* (pp. 1–10). New York, NY: IEEE.

Harle, R. (2013). A survey of indoor inertial positioning systems for pedestrians. *IEEE Communications Surveys and Tutorials, 15*(3), 1281–1293.

Helbing, D. (1992). Models for pedestrian behavior. *Natural structures: Principles, strat-egies, and models in architecture and nature, Part II (Sonderforschungsbereich 230, Stuttgart, 1992)*, pp. 93–98.

Helbing, D., & Molnar, P. (1995). Social force model for pedestrian dynamics. *Physical Review, E 51*, 4282–4286.

Howe, J. (2006). The rise of crowdsourcing. *Wired Magazine, 14*(6).

Hyytiä, E., Lassila, P., & Virtamo, J. (2006, June). Spatial node distribution of the ran-dom waypoint mobility model with applications. *IEEE Transactions on Mobile Computing, 5*(6), 680–694.

Ibach, P., Stantchev, V., Lederer, F., Weiß, A., Herbst, T., & Kunze, T. (2005). WLAN-based asset tracking for warehouse management. In *Proceedings of the International Conference e-Commerce, 2005*. Porto, Portugal.

Kaiser, S., Garcia Puyol, M., & Robertson, P. (2012). Maps-based angular PDFs used as prior maps for FootSLAM. In Proceedings of the IEEE/ION *Position location and navigation symposium (PLANS), 2012*. New York, NY: IEEE.

Kaiser, S., Garcia Puyol, M., & Robertson, P. (2013). Measuring the uncertainty of probabilistic maps representing human motion for indoor navigation. *IEEE Transactions on Information Theory*, 1–29.

Kaiser, S., Khider, M., & Robertson, P. (2011). A human motion model based on maps for navigation systems. *EURASIP Journal on Wireless Communications and Networking, 60*.

Kaiser, S., Khider, M., & Robertson, P. (2012). A pedestrian navigation system using a map-based angular motion model for indoor and outdoor environments. *Journal of Location Based Services, 7*(1), 44–63.

Kammann, J., Angermann, M., & Lami, B. (2003). A new mobility model based on maps. In *Proceedings of the Vehicular technology conference VTC, 2003*. New York, NY: IEEE.

Khider, M., Kaiser, S., & Robertson, P. (2012). A novel 3-dimensional movement model for pedestrian navigation. *Journal of Navigation, 65*(2), 245–264.

Khider, M., Kaiser, S., Robertson, P., & Angermann, M. (2008). A novel movement model for pedestrians suitable for personal navigation. In *Proceedings of the national technical meeting of the Institute of Navigation ION NTM 2008*. Manassas, VA: Institute of Navigation.

Khider, M., Kaiser, S., Robertson, P., & Angermann, M. (2009). Maps and floor plans enhanced 3D movement model for pedestrian navigation. In *Proceedings of the 22nd international technical meeting of the satellite division of the Institute of Navigation. 2009*. Manassas, VA: Institute of Navigation.

Klingbeil, L., Romanovas, M., Schneider, P., Traechtler, M., & Manoli, Y. (2010). A mod-ular and mobile system for indoor localization. In *Proceedings of the International conference on indoor positioning and indoor navigation (IPIN)*. New York, NY: IEEE.

Kotanen, A., Hannikainen, M., Leppakoski, H., & Hamalainen, T. (2003). Positioning with IEEE 802.11b wireless LAN. In *Proceedings of 14th IEEE conference on per-sonal, indoor and mobile radio communications, PIMRC 2003*. New York, NY: IEEE.

Krach, B., & Robertson, P. (2008). Cascaded estimation architecture for integration of foot-mounted inertial sensors. In *Proceedings of the IEEE/ION position location and navigation symposium*. New York, NY: IEEE.

Krumm, J. (2008, April). A Markov model for driver turn prediction. *SAE SP* .

Lakoba, T. I., Kaup, D. J., & Finkelstein, N. M. (2005). Modifications of the Helbing-Molnár-Farkas-Vicsek social force model for pedestrian evolution. *SIMULATION, 81*(5), 339–352.

Liu, H., Darabi, H., Banerjee, P., & Liu, J. (2007). Survey of wireless indoor positioning techniques and systems. In *IEEE transactions on systems, man, and cybernetics, Part C: Applications and reviews*. New York, NY: IEEE.

Mirowski, P., Ho, T. K., Yi, S., & MacDonald, M. (2013). SignalSLAM: Simultaneous localization and mapping with mixed WiFi, Bluetooth, LTE and magnetic signals. In *Proceedings of the Indoor positioning and indoor navigation (IPIN), 2013*. New York, NY: IEEE.

Montemerlo, M., Thrun, S., Koller, D., & Wegbreit, B. (2002, July). FastSLAM: A factored solution to the simultaneous localization and mapping problem. *AAAI/ IAAI*, 593–598.

Nilsson, J.-O., Skog, I., & Händel, P. (2010). Performance characterisation of foot-mounted ZUPT-aided INSs and other related systems. In *Proceedings of the International conference on indoor positioning and indoor navigation (IPIN), 2010*. New York, NY: IEEE.

Okazakia, S., & Matsushitaa, S. (1993). A study of simulation model for pedestrian movement with evacuation and queuing. In *Proceedings of the international conference on engineering for crowd safety*.

Osorio, C., & Michel Bierlaire, M. (1993). *An analytic finite capacity queuing network capturing congestion and spillbacks*. Tristan VI, EPFL.

Parsons, J. (2000). *The mobile radio propagation channel* (2nd ed.). Hoboken, NJ: John Wiley & Sons.

Pietrzyk, M., & von der Grün, T. (2010). Experimental validation of a TOA UWB ranging platform with the energy detection receiver. In *Proceedings of the IEEE International conference on indoor positioning and indoor navigation (IPIN), 2010*. New York, NY: IEEE.

Ressel, W. (2004). Modeling and simulation of mobility. In *Proceedings of the 1st international workshop on intelligent transportation (WIT, 2004)*. Hamburg, Germany: Technical University of Hamburg-Harburg.

Ristic, B., Arulampalam, S., & Gordon, N. (2009). *Beyond the Kalman filter: Particle filters for tracking applications*. Boston, MA, and London, UK: Artech House.

Robertson, P., Angermann, M., & Krach, B. (2009). Simultaneous localization and mapping for pedestrians using only foot-mounted inertial sensors. In *Proceedings of the 11th international conference on ubiquitous computing*. New York, NY: ACM.

Robertson, P., Frassl, M., Angermann, M., Doniec, M., Julian, B. J., Puyol, M. G. (2013). Simultaneous localization and mapping for pedestrians using distortions of the local magnetic field intensity in large indoor environments. In *Proceedings of the IEEE international Conference on Indoor positioning and indoor navigation (IPIN), 2013*. New York, NY: IEEE.

Robertson, P., Garcia Puyol, M., & Angermann, M. (2011). Collaborative pedestrian mapping of buildings using inertial sensors and footslam. In *Proceedings of the 24th international technical meeting of the satellite division of the Institute of Navigation*. Manassas, VA: Institute of Navigation.

Schmidt, G. K., & Azam, K. (1993). Mobile robot path planning and execution based on a diffusion equation strategy. *Advanced Robotics, 7*(5), 479–490.

Sharma, S., & Vishwamittar. (2005, August). Brownian motion problem: Random walk and beyond. *Journal of Science Education: Resonance,* 49–66.

Shen, G., Chen, Z., Zhang, P., Moscibroda, T., & Zhang, Y. (2013). Walkie-Markie: Indoor pathway mapping made easy. In *Proceedings of the 10th USENIX conference on networked systems design and implementation.* Berkeley, CA: USENIX Association.

Shlesinger, M. F., Zaslavskii, G. M., & Frisch, U. (1994). *Lévy flights and related topics in physics: Proceedings of the international workshop held at Nice, France.* Berlin, Germany. Springer.

Skog, I., Nilsson, J.-O., & Händel, P. (2010). Evaluation of zero-velocity detectors for foot-mounted inertial navigation systems. In *Proceedings of the IEEE International conference on indoor positioning and indoor navigation (IPIN).* New York, NY: IEEE.

Soloviev, A., & Miller, M. (2010). Navigation in difficult environments: Multisensor fusion techniques. In *Proceedings of NATO lecture series under the sponsorship of the Sensors and Electronics Technology Panel (SET).* New York, NY: Springer.

Staudinger, E., Klein, C., & Sand, S. (2011). A generic OFDM based TDoA positioning testbed with interference mitigation for subsample delay estimation. In *Proceedings of 8th international workshop on multi-carrier systems and solutions.* New York, NY: IEEE.

Teknomo, K. (2002). *Microscopic pedestrian flow characteristics: Development of an image processing data collection and simulation model* (Unpublished PhD dissertation). Tohoku University, Japan.

Weifeng, F., Lizhong, Y., & Fan, W. (2003). Simulation of bi-direction pedestrian movement using a cellular automata model. *Physica A: Statistical Mechanics and Its Applications, 321,* 633–640.

Wendlandt, K., Robertson, P., Khider, M., Angermann, M., & Sukchaya, K. (2007). Demonstration of a realtime active-tag RFID, Java based indoor localization system using particle filtering. In *Adjunct Proceedings of UbiComp, 2007.*

Weyn, M., & Schrooyen, F. (2008). A WiFi-assisted GPS positioning concept. In *Proceeding of the third European conference on the use of modern information and communication technologies.*

Woodman, O. (2007). *An introduction to inertial navigation.* Cambridge, UK: University of Cambridge, Computer Laboratory.

Woodman, O., & Harle, R. (2008). Pedestrian localisation for indoor environments. In *Proceedings of the 10th international conference on ubiquitous computing* (pp. 114–123). Seoul, South Korea: ACM.

Yang, L., Fang, W., Li, L., Huang, R., & Fan, W. (2003). Cellular automata pedestrian movement model considering human behavior. *Chinese Science Bulletin, 48*(16), 1695–1699.

Zampella, F., Khider, M., Robertson, P., & Jiménez, A. (2012). Unscented Kalman filter and magnetic angular rate update (MARU) for an improved pedestrian dead-reckoning. In *Proceedings of the IEEE/ION Position location and navigation symposium (PLANS), 2012 IEEE/ION.* New York, NY: IEEE.

Zarimpas, V., Honary, B., Lund, D., Tanriover, C., & Thanopoulos, N. (2005). Location determination and tracking using radio beacons. In *2005, Proceedings of the 6th IEE international conference on 3G and beyond.* United Kingdom: IET.

# 7

# Indoor Navigation Challenges for Visually Impaired People

M. Beatrice Dias

Ermine A. Teves

George J. Zimmerman

Hend K. Gedawy

Sarah M. Belousov

M. Bernardine Dias

## CONTENTS

*Abstract:* The ability to independently navigate urban environments is a fundamental necessity for all of us. While outdoor navigation is a long-studied and well-established area of research that has yielded many practical solutions, analogous solutions for navigating indoor environments are still relatively sparse. The problem of indoor navigation is further complicated for those who have visual impairments. Visual impairments can range from partial blindness to severe visual impairment. People who are visually impaired must use a variety of techniques to familiarize themselves with new environments, to orient themselves within an unfamiliar environment, and to navigate an environment to move between points of interest. While white canes and guide dogs can assist people who are visually impaired with obstacle avoidance and some aspects of safe passage, many of the challenges faced by these individuals when navigating unfamiliar indoor environments remain unsolved. Perhaps the most critical need is for tools that can alert visually impaired people during dangerous situations and safely guide them during emergency evacuations. As a result of these challenges, people with visual impairments have to rely on sighted people when they need to visit an unfamiliar location and are often limited to visiting a few familiar places.

    From a technical perspective, solutions for people with visual impairments require a higher level of accuracy in terms of localization and obstacle or hazard avoidance, customized path-planning algorithms that take into account preferences and landmarks accessible to visually impaired people, and user interfaces that are both accessible and customizable. To develop mobility aids or devices that most effectively serve the navigation needs and preferences of people who are visually impaired, researchers and technologists must better understand the nature, scope, complexity, and diversity of the challenges faced by people with visual impairments and the current navigation methods they employ. Toward this end, this chapter outlines the needs and challenges for indoor wayfinding and navigation faced by individuals who are visually impaired. This chapter is based on findings from several years of needs assessment conducted with relevant experts and people who are visually impaired.

## 7.1 Introduction

Imagine trying to navigate an unfamiliar building without the aid of signs to indicate where the exits are, or notices to let you know where the elevators and stairs are, or warning posts to alert you about a wet floor or a construction area. More critically, what happens when you cannot follow the signs to the nearest exit in case of an emergency or do not even know that you are in a dangerous situation and should evacuate a building? To learn about the general spatial layout of an unfamiliar building or the best routes to exit a building in case of an emergency evacuation, you might ask a friend or bystander for assistance, or you might refer to a two-dimensional map for reference. Now imagine that you have entered this unfamiliar building, but you are visually impaired. How would you go about obtaining the information you need to travel safely and confidently to reach your destination within this unfamiliar indoor environment? People who are visually impaired are confronted with wayfinding and navigation problems such as these on a daily basis. Given such challenges, individuals who are visually impaired are often unable to explore new spaces on their own and typically have to rely on friends, family members, hired experts, or others when they need to navigate an unfamiliar location. This severely limits the number of unfamiliar places they can visit.

We used several approaches to understand the processes, techniques, tools, challenges, and unmet needs of visually impaired people relevant to wayfinding and navigation. These methods include passive observations, examination of relevant online blog posts, interviews, and surveys. These diverse approaches yielded a significant set of data that should inform the development of more effective and useful navigation tools for travelers who are visually impaired. Furthermore, toward ensuring a successful needs assessment process, we initiated a working relationship with the Blind & Vision Rehabilitation Services of Pittsburgh (BVRSP)* and the Western Pennsylvania School for Blind Children (WPSBC).† These organizations and their networks primarily informed the needs disseminated in this work.

BVRSP is a private, nonprofit, United Way agency that believes in independence through rehabilitation. For over 100 years, BVRSP has worked with people who are blind, deaf-blind, or vision impaired to help them become independent. BVRSP instructors, many of whom are blind or vision impaired themselves, provide instruction in essential areas. They also offer vocational and employment services, a low-vision rehabilitation program, and employment. We worked with several blind or visually impaired adults from BVRSP and their network to inform this work.

---

* Blind & Vision Rehabilitation Services of Pittsburgh: http://bvrspittsburgh.org/.
† Western Pennsylvania School for Blind Children: http://www.wpsbc.org/.

Founded more than 120 years ago, WPSBC is an educational facility committed to educating students who are visually impaired with severe concomitant disabilities. The majority of their students have cognitive and ambulatory challenges, and the school's facility and programming are tailored for boys and girls who require distinct educational and supportive services. The curriculum emphasizes the acquisition of life skills. WPSBC also has a large network of alumni, many of whom are active in organizations in Pittsburgh. We benefited from the experience and expertise of the WPSBC staff in carrying out this work.

Sighted individuals take for granted their ability to walk into a building and find their way around independently. In contrast, individuals who are visually impaired must use a variety of techniques to orient themselves within an unfamiliar environment, determine a suitable pathway to their destination, and navigate the physical space to move between points of interest. White canes and guide dogs can assist with obstacle avoidance and some aspects of safe passage, but many of the challenges faced by individuals with visual impairment when navigating unfamiliar environments remain unsolved. Perhaps the most critical need is for techniques to allow this population to be meaningfully alerted of dangerous situations and safely guided during emergency evacuations indoors. This chapter explores the challenges encountered and techniques employed by visually impaired people navigating unfamiliar indoor spaces, with the intention of informing researchers and technologists about these needs and challenges so that they can develop more effective navigational aids and associated technology tools for the visually impaired population.

## 7.2 Orientation and Mobility for People Who Are Blind or Visually Impaired

Wayfinding and navigation through familiar and unfamiliar indoor or outdoor spaces for individuals who are blind or visually impaired involve a combination of cognitive and sensory processing (for orientation) and constant alertness for hazards (for safe mobility). Because of their inability to use proximal and distal visual cues, blind or visually impaired individuals use alternate sensory input (auditory, tactile, proprioception) and strategies to safely orient and navigate within a space. Some of these strategies include identifying landmarks and clues to remain oriented. A landmark is defined as any unique single datum or multiple sensory data consistently experienced in a given environment. So a single sensory experience, such as a change in kinesthesia or proprioception caused by an uprooted slab of concrete, can be used to identify that specific location in the overall outdoor environment, but that landmark must be unique and it must be experienced each time the individual travels that route. A change in floor texture from carpet to tile is an example of an indoor landmark. A clue is any sensory information that is not unique or consistently experienced in a given environment. For example, while the smells from a coffee shop are unique to a specific environment, the traveler may not perceive those smells because of weather conditions or the shop being closed. Similarly, in the example of the uprooted slab, there may be more than one such slab on a given sidewalk, in which case that sensory experience would not be unique. The issues of wayfinding in novel outdoor environments in the population of individuals who are blind or visually impaired have been well documented in the literature (Golledge, Marston, Loomis, & Klatzky, 2004; Long & Giudice, 2010; Zimmerman, 1990).

In addition to the use of landmarks and clues for orientation, other strategies include recall of mental or cognitive maps of familiar places, systematic familiarization of unfamiliar places, and soliciting assistance (physical or verbal) from the public to obtain location information. Blind or visually impaired individuals use these strategies as they navigate through space using one or more mobility tools for protection and identification. Most common is the use of a human guide. The guide leads the individual through space, avoiding potential hazards to reach the destination. The downside to this method is dependency, because the blind or visually impaired person must depend on someone else for assistance. The long white cane is another common mobility tool most associated with individuals who are blind or visually impaired. The traveler holds the cane in front and two steps ahead of the body, while moving it side to side, spanning the width of his or her body to probe the environment, noting textural and elevational changes on the walking surface. It also tactually informs the user about objects (e.g., furniture) in his or her path. A dog guide is a highly specialized but effective mobility tool that is used by a small percentage of individuals who are blind or visually impaired. Guide dogs are primarily used by young adults (beginning at age 16 years) or adults. In addition to these tools, other devices such as handheld GPS devices or monocular distance telescopes may be used in combination with one of the previously mentioned mobility tools for outdoor navigation. The orientation and mobility (O&M) specialist is the professional who is responsible for teaching these cognitive and sensory strategies and the use of all of the mobility tools and devices to individuals who are blind or visually impaired. The O&M specialist teaches skills of safe and independent travel.

### 7.2.1 History

O&M instruction began during World War II when blinded soldiers were sent to three veteran administration hospitals across the United States to receive medical and rehabilitation attention for their war-related injuries. The federal government decided that rehabilitation of blinded soldiers should become a priority (which was the first time that rehabilitation of service personnel became a part of the role of the federal government) to returning the war wounded to civilian life, so a new program of independent travel using a white cane and human guide skills was developed. That program is the basis of the current O&M skills used to instruct children and adults who are blind or visually impaired today.

### 7.2.2 Sequence of Instruction and Familiarization

O&M is based on a developmental curriculum model of instruction. Less complex skills of human guide and various other indoor safety skills and techniques (such as hand trailing a wall) are introduced before more advanced complex outdoor skills and techniques are introduced (e.g., subway travel). That said, even the most adept travelers have a need for indoor and/or outdoor familiarization services from an O&M specialist. The standard O&M familiarization service includes an initial human-guided walkthrough of the route being learned, pointing out various sensory landmarks or clues that are particularly useful for orientation purposes. Depending on the traveler's ability and/or the complexity of the route being learned, there may need to be multiple guided walks, but each walk would allow the traveler the opportunity to establish the time, distance, and rate for the entire route, as well as to identify relevant landmarks critical for determining orientation. The O&M specialist may use a tactile map to provide the traveler with an aerial perspective of the entire spatial layout of the environment, as well as the specific route(s) within that environment. Once the traveler feels more comfortable with his or her spatial orientation to the route, the O&M specialist will take the traveler to the starting point and allow the traveler to use the white cane or dog guide (whatever the traveler is using at the time), while the O&M specialist walks along in close proximity, auditorially reinforcing the sensory landmarks and clues. The O&M specialist's proximity to the traveler, especially at complex street crossings, is critical during this phase of familiarization. Finally, the O&M specialist will fade back a distance from the traveler and allow the individual to travel the route "independently," intervening when necessary.

### 7.2.3 Technology and O&M

GPS technology is beginning to change how some people who are blind or visually impaired acquire outdoor spatial orientation knowledge, but GPS technology has limited use indoors, putting the traveler in the position of depending on human assistance to locate a classroom or office. Karimi, Nwan, and Zimmerman (2009) discussed the use of model-centric technology versus experience-centric technology for navigation. In the model-centric approach, the process to compute a particular route from one point to another may not be any more effective than using a friend or an unskilled individual, since it would not take risk into account. An O&M specialist, whose professional knowledge and application is based on the experience-centric approach, would look at the potential routes from one location to another and select the safest (lowest risk to client), most efficient, or shortest route (e.g., a route that would involve a complex offset intersection with multiple lanes of traffic turning in various directions would most likely be avoided). Alternatively, the model-centric approach may include a complicated traffic intersection in a recommended travel route, which in turn could place the individual in a possibly life-threatening situation. A better approach is to tailor the routes (e.g., experience-centric) from each starting point and location through the lens of the O&M specialist.

## 7.3 Related Work

There is a growing literature on technology solutions for blind and visually impaired users with regard to indoor wayfinding and navigation. However, there is a shortage of usability studies that investigate the point of view of this population and understand users' needs when it comes to navigation support and solution design (Sánchez & Elías, 2007). This section summarizes findings of studies that conducted needs assessment with blind and visually impaired users to inform and guide the design of navigational technology solutions.

Golledge et al. (2004) surveyed 30 participants who were blind or visually impaired to learn more about their preferences for a possible navigation device. Their findings highlight that the need for preview of the novel environment to gather information before making a walking trip to an unfamiliar area is critical to independent travel within that environment. Information such as landmarks, route information, and building information is essential to creating a mental map of the unfamiliar destination. The authors further stated that individuals who are blind or visually impaired prefer speech input and output interfaces in a technology navigation solution. Finally, the authors noted that the output should not interfere with the environmental clues necessary for navigation.

In another study by Sánchez and Elías (2007), researchers learned that white cane users encounter environmental obstacles, such as arches, beams, and chandeliers, that are above waist height and are not easily detected by the cane alone. According to the researchers, while individuals who are blind or visually impaired want to gather information about their surroundings, they do not necessarily want others to tell them about everything. The study advocates presenting only a small subset of the available information to visually impaired travelers in accordance with their needs during navigation. This suggests that technology aids should allow the users to solve problems and make decisions on their own, thus encouraging the development and use of the O&M skills learned and their own abilities. In other words, technologies should assist and enhance the skills of the user rather than make the user reliant on them.

Quiñones, Greene, Yang, and Newman (2011) interviewed 20 individuals who were blind or visually impaired and learned that even in familiar routes, changes in the environment can derail their wayfinding process. Challenges the participants encountered were loss of clues, obstacles, and environmental conditions. Many people ask for help when lost, but the traveler cannot always count on someone being available and capable of providing accurate directions.

A more recent study by Williams, Hurst, and Kane (2013), based on interviews with 30 adults with vision impairments, provided more insight into the variety of behaviors and attitudes of blind and visually impaired people with regard to navigation and technology. All participants received at least some O&M training and specified their primary mobility navigation aid (e.g., white canes, guide dogs, and human guides). In situations where navigation technology failed, they always relied on their O&M skills. This highlights the importance of technology solutions to supplement and augment, rather than replace, existing training techniques. Furthermore, depending on a person's primary mobility navigation aid, he or she tends to treat obstacles differently. For example, cane users utilize their aid to detect obstacles, whereas dog guide users utilize their aid to avoid obstacles. People who are visually impaired also rely on human guides, especially in indoor environments, since these locations rarely offer accessible maps, and layouts change frequently. When people navigate public spaces, interacting with strangers is inevitable. Sometimes these interactions can be stressful or dangerous, and many participants were concerned about exposing expensive technological devices in public. This highlights the importance of a multimodal interface for a navigation technology solution, so that a visually impaired person can decide the appropriate input and output in a given situation.

All of the aforementioned studies also discuss the cosmetic acceptability of such a navigational device. Most people would find a wearable navigational device to be acceptable; however, it should not interfere with the primary mobility navigation aid they use (Golledge et al., 2004; Sánchez & Elías, 2007; Williams et al., 2013). Furthermore, affordability and the desire to carry as few tools as possible are factors that researchers should take into consideration when designing such solutions (Quiñones et al., 2011; Williams et al., 2013).

Findings from the literature are in line with the outcomes of our needs assessment efforts, which further enhance these conclusions and provide additional details. The results of our needs assessment are described in more detail later in this chapter.

## 7.4 Needs Assessment Methods

To inform solutions that most effectively serve the navigation needs and preferences of individuals with visual impairment, we conducted a needs assessment study to gain a better understanding of the nature, scope, complexity, and diversity of their current navigation methods and challenges. While examining existing navigation solutions, we also gathered information and feedback from partner organizations to inform and guide the work.

### 7.4.1 Participants

The research population included participants who are visually impaired, O&M specialists, and building managers based in Pittsburgh, Pennsylvania, United States. Following Carnegie Mellon University's Institutional Review Board (IRB) procedures, a protocol was set up and approved for this research project, and participants were recruited with help from contacts within the partner organizations. In total, the pool of participants included 20 individuals who are visually impaired, 7 O&M specialists, and 5 building managers.

### 7.4.2 Data Collection

We employed a combination of surveys, interviews, observations, and secondary data to enhance the understanding of the navigation challenges and tools of visually impaired travelers. The needs assessment was conducted in five phases, which are described next. To preserve the privacy of research participants, all survey and interview data were recorded anonymously.

#### 7.4.2.1 Phase 1

- *Passive observations:* Researchers observed indoor navigation lessons for students and a fire drill evacuation process at one of our partner locations. The navigation lessons consisted of two hours of passive observation of the two activities inside the main school building: (1) navigation training session for a person with limited vision who also uses a wheelchair and (2) navigation training session with a young child who is totally blind but is able to move without a wheelchair and uses a white cane to navigate.

- *Initial interviews:* Researchers interviewed three staff members who are visually impaired to learn about their navigation experiences and their interaction with technologies. Four O&M specialists were interviewed to learn more about what instructions they use to describe routes to travelers who are visually impaired. During the interviews, researchers also asked several questions related to navigation needs and interface input-output modalities that work for the target user. These initial interviews focused on learning about the following topics:

- *User interfaces of successful technological interventions adopted by the visually impaired community:* Researchers concentrated on those interventions that could be directly related to the navigation domain or mobile devices.

- *Learning and familiarization processes related to the use of mobile technology:* Researchers focused on understanding experiences that provided clues about how users who are visually impaired adopt mobile navigation aids.

- *Types of interfaces preferred or rejected by users who are visually impaired and were already familiar with mobile devices:* This aspect included input and output modalities applicable to navigation systems, as well as features related only to the information given by navigation aids (e.g., level of verbosity).

- *Types of activity for which a user who is visually impaired would like to get navigation assistance* (e.g., shopping, emergency evacuation).

- *Impediments for the adoption of electronic travel aids* (e.g., trust levels in current technology, fears about being lost or confused, independence, and lack of motivation).

### 7.4.2.2 Phase 2

- *Examination of online blog posts:* To understand the extreme cases of visual impairment, the research team read online blogs written by individuals who are deaf-blind. While written from the perspective of the deaf-blind population, most of the information still applied generally to individuals with visual impairment only. Entries included information about what individuals who are deaf-blind pay attention to during navigation, such as landmarks and environmental clues. The blog entries also provided some insight into some of their challenges.

- *Examination of instruction models and narrative maps:* A variety of instruction models are used with existing navigation technologies. Researchers examined these resources to understand the types of guidance offered to travelers who are visually impaired.

### 7.4.2.3 Phase 3

This phase included detailed surveys and interviews with 20 participants from partner organizations. Surveys were conducted online, and interviews were conducted either over the phone or online in a survey format. Of the participants, 18 were individuals who were visually impaired, and the remaining 2 participants had no impairments but worked closely with visually impaired individuals. The goal of this phase was to learn more about end users' navigation and localization experiences, as well as to identify features that are most important to target users of assistive navigation aids.

### 7.4.2.4 Phase 4

The final phase of needs assessment consisted of interviews with building managers to understand their perspective in enabling visually impaired people to safely navigate the indoor environments they manage. Five managers were included to understand their challenges in making buildings accessible to visitors who are visually impaired. These five individuals managed buildings for a mix of for-profit and nonprofit organizations and offered a different perspective of the difficulties that can arise in making indoor environments accessible to visitors with visual impairments.

## 7.5 Needs Assessment Findings

The findings of our needs assessment study are detailed next. Where relevant, we comment on how our findings inform technology solution designs for empowering visually impaired travelers to safely and independently navigate indoor environments.

### 7.5.1 Indoor Orientation and Location Identification Techniques

The majority of participants reported that landmarks (e.g., doorways, elevators) and environmental clues (e.g., smells, sounds) are very important in facilitating their indoor navigation, particularly in terms of orienting themselves within a building. In particular, an examination of blog posts by deaf-blind individuals during Phase 2 highlighted the use of landmarks and environmental clues to inform their navigation (see Table 7.1). In addition, 60% of Phase 3 participants reported that they used environmental clues to orient themselves, with 55% of these individuals preferring this method of orientation over any other. Frequently used landmarks and environmental clues are listed in Table 7.2.

### 7.5.2 Indoor Navigation Techniques

The way that an individual who is visually impaired navigates may depend on when that person lost vision. However, two salient navigation techniques emerged from our needs assessment efforts:

**TABLE 7.1**

A Sample of Relevant Blog Quotes from Phase 2

"I moved around a couple corners, down the hall, past two doors.... I kept my hand on the wall so I would know what I was passing" (Orlando, 2009b).

"Unfortunately I wasn't able to mentally map the layout of the building. I was too busy trying to find a familiar landmark" (Orlando, 2009b).

"I realized I was not where I should be. The approach was wrong. So was the angle of the door handle" (Orlando, 2009a).

"The rubber mat felt right.... I followed the edge of the mat with my cane.... But the mat ended too soon" (Orlando, 2009a).

"The feel of the rubber mat tells me that I am nearing the bulletin board and need to be ready to cross the hall" (Orlando, 2009a).

"Every time I go to class, I walk down that hall and pass that location. And every time I do, I smell coffee right before I hit the rubber mat" (Orlando, 2009a).

"I know the campus does deep cleaning during breaks.... The halls can be a mess. It makes it extra hard for me to get around.... The mess covered up all my landmarks" (Orlando, 2010a).

"But the wind was also interfering with my ability to use scent and touch.... Because of the constant wind against my skin, I couldn't feel the displacement of air as people moved past me" (Orlando, 2010b).

"I'm mostly deaf. I only hear environmental sounds with my old cochlear implant. I can't understand speech.... I can hear the chatter of people, the rustling of papers, and the sound of doors being slammed shut. Or I can hear the silence" (Orlando, 2010b).

"I can smell the scent of people.... Sometimes I smell food, as someone eats a snack near me. I can often determine my location by scent, as well. Hallways smell bland and stale. I can smell coffee near the snack room" (Orlando, 2010b).

"Touch and the displacement of air give me more useful clues. There is a slight movement in the air when people walk past.... Or I can feel a 'whoosh' of cold or fresh air when someone opens the building doors" (Orlando, 2010b).

**TABLE 7.2**

Environmental Clues and Landmarks Used by Visually Impaired People When Navigating Indoor Spaces

| Environmental Clues | Indoor Landmarks |
| --- | --- |
| • Sounds | • Bulletin boards |
| • Smells (e.g., coffee, food) | • Intersections |
| • Floor contrasts | • Turns |
| • Hallway length | • Doors and doorways |
| • Size and feel of floor mats | • Vending machines |
| • Feel of displacement of air | • Water fountains |
| • Wall markings | • Elevators |
| • Braille and raised signs | • Stairwells |
| • Size of steps | • Information or front desks |
| | • Fire alarms |
| | • Walls |
| | • Corners |
| | • Tables |

- *Making a mental or cognitive map of the building layout:* Observations at the WPSBC highlighted the important role that mental or cognitive maps play when training students how to navigate within the school premises. In fact, most needs assessment participants reported that creating a conceptual layout of a building they frequently visit or plan to visit is extremely useful.
- *Counting environmental clues and landmarks (e.g., corners and doors):* Of Phase 3 respondents, 70% used a landmark counting method (e.g., counting the number of doorways or intersections) to aid them with navigation.

In addition, to preplan their travel, participants utilize relevant websites and applicable smartphone apps, call the destination location prior to their departure, or request help from a friend. In cases where these techniques are insufficient to guide them to their destination, travelers who are visually impaired rely on assistance from sighted peers or bystanders. However, sighted people are not always aware of how an individual who is visually impaired needs to receive directions to get to a destination since they typically only think of ways a sighted person would visualize a route. Therefore, requesting assistance from the sighted public can also be challenging. For example, a traveler who is visually impaired needs to receive directions such as "walk until you hear the sound or feel the wind of the intersecting hallway, then turn right." These directions do not necessarily have to be that detailed, but if there are any navigational or sensory clues that the person who is giving directions could think of to make it concrete and meaningful, then that alone would be very helpful.

Apart from these practices, the process of navigation may vary among travelers who are visually impaired. In particular, the speed of movement and frequency of orientation differed across Phase 3 participants, as depicted in Figure 7.1.

### 7.5.3 Wayfinding Maps

The ClickAndGo Wayfinding Maps service provides very detailed and high-quality narrative maps for indoor and outdoor routes.* We reviewed 20 ClickAndGo indoor narrative map instructions. A narrative map is an auditory or text-based description that provides wayfinding instructions for following a walking route and maintaining orientation along the route. These instructions are manually prepared by specialists. The founder of the ClickAndGo service typically traverses each route in person, videotapes it, and records routing directions for it. The instructions can then be downloaded from the website in text format. This format can be used in devices that support Text-To-Speech and Text-To-Braille. The instructions can also be

---

* ClickAndGo Wayfinding Maps: http://www.clickandgomaps.com/narrative-maps.

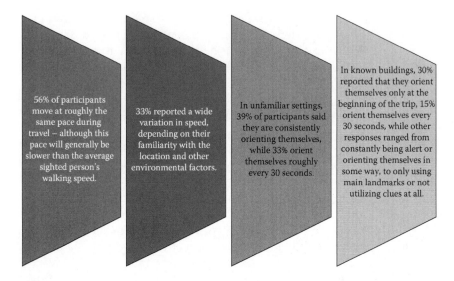

**FIGURE 7.1**
Phase 3 participants' varying speed of movement and frequency of orientation.

downloaded as MP3 files that are compatible with portable audio devices. There is also a voice service where the user can call free of charge to ask for routing directions from one location to another.

Manually creating these routes takes a lot of time and effort, making it difficult to scale this system to many places or buildings. However, the ClickAndGo advantage over other services or technologies is the quality of the instructions it provides. On the basis of testimonials provided on the ClickAndGo website, users find the instructions outstanding and provide very good environmental clues.

### 7.5.4 Technology Use

In familiar settings, most respondents navigate independently without too many challenges. However, when they are lost in a building or navigating an unfamiliar indoor space, asking for help is the most popular strategy they use to find their way. If unassisted and disoriented in an unfamiliar environment, a person with a visual impairment faces a significant challenge navigating that space and may even be at risk of venturing into hazardous areas of that building. This demonstrates a need for navigation assistance, particularly in unfamiliar locations. Technology can help fill this need. Our survey of participants revealed an array of technologies they currently use to assist with their travel (see Figure 7.2). Although these tools facilitate their wayfinding and navigation in many ways, there is still a need for improved and more comprehensive technologies to better address the needs and preferences of travelers who are visually impaired.

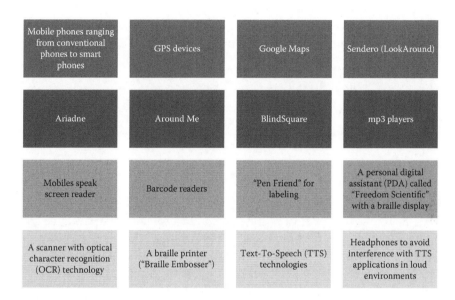

**FIGURE 7.2**
Assistive navigation technologies currently used by survey participants.

Most participants who are visually impaired had a relatively high level of comfort interacting with technology. In particular, all respondents from Phase 3 use electronic devices, including a computer, digital readers, scanners, CCTVs, notetakers, and GPS. These participants feel comfortable using these tools, navigating technology menus, and listening to and following instructions to operate technology applications. Of Phase 3 participants, 60% had smartphones, while 35% had feature phones, and the remainder did not report the use of a cell phone. Also notable is that 50% of participants who had smartphones had an iPhone, while none reported using an Android phone. However, one iPhone user mentioned feeling quite limited and would like to "make the jump" to an Android, suggesting that the Android may have many appealing features for users who are blind. The majority of respondents were comfortable using their phones for many different purposes, and more than half of these individuals used their phones to assist with some aspect of navigation. Fewer than 40% of participants indicated that they use their phones exclusively for phone calls.

The need for technology training varies greatly depending on the individual and device in question. Of Phase 3 participants, 78% required multiple days to learn more complex technologies (several days to weeks or longer). On the other hand, 72% said they could learn more simple technologies in less than a day. Although 33% of participants typically use personal training or assistance with new technologies, 61% do not. Respondents also reported that they learn new technologies most rapidly when the directions and lay-out are clear and intuitive and when they have access to braille instructions.

### 7.5.5 Technology Preferences

Information about technology preferences for the visually impaired are derived from Phase 1 and Phase 3 of this study and are presented next.

#### 7.5.5.1 Technology Input Mode

The most popular method of technology input was through speech or voice commands. Tactile buttons and touch screens were also rated highly among participants. Although most participants indicated that they would be comfortable drawing gestures on a touch screen, one respondent indicated that it might be difficult knowing where to draw the gestures. Respondents also mentioned the use of a screen reader. Thus, current phone use reveals a preference for speech-enabled features, although a few respondents reported experiencing problems with speech input and voice recognition, particularly in crowded and noisy areas. Figure 7.3 depicts participant preferences for input modalities.

#### 7.5.5.2 Technology Output Mode

Audio output was the most preferred technology output mode. However, participants voiced concern that audio output from a navigation device could hinder their ability to recognize environmental sound clues, which most of them pay attention to while traveling. Therefore, although they recognize the importance of audio output, it is vital that users are able to adjust and customize sound output and verbosity according to their surroundings, as well as their individual needs and preferences. Nevertheless, 50% of Phase 3 participants did not think audio output from a device would interfere with their navigation.

Some respondents preferred a combination of vibration and audio output, and others preferred a combination of vibration, audio, and braille. These

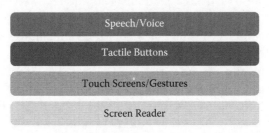

**FIGURE 7.3**
Preferred technology input modalities.

results indicate a preference for a device that could provide both vibratory and audio feedback and also include options to turn audio or vibrations on and off as preferred or needed. In terms of potential headphone use with the device, most participants utilize both speakers and headphones with their devices, depending on the situation. However, several commented that headphones could be used with a navigation device only if used in just one ear, because they cannot be totally cut off from environmental sound clues.

### 7.5.5.3 Customizations

The majority of respondents thought that there should be different modes of instruction based on familiarity of the environment. Furthermore, different participants highlighted the value of using context-sensitive information. For example, respondents wanted to know where the elevators are located, where the front desk is, what the sizes of the steps are, and so on. In addition, participants wanted the capability to choose the level of detail and types of information they receive from their navigation technology. In terms of receiving directions and instructions from their navigation aid, some respondents preferred a combination of high-level, periodic, and/or on-demand instructions. Others wanted just on-demand, periodic, or high-level instructions. These responses highlight the need for customizable assistive navigation tools. Respondents also commented on features they like and dislike in currently available technologies, which additionally emphasized the need for customizability and user control when feasible. Table 7.3 lists these preferences.

### 7.5.6 Barriers to Technology Adoption

Different barriers, including price, portability, and how much they block the users' other senses, make the adoption of electronic travel aids difficult. One participant noted that data plans for the latest generation of mobile devices are expensive, while another emphasized the high cost of GPS and color identification devices. An interesting issue raised by all was the lack of standardization across navigation applications. A poor level of user friendliness

**TABLE 7.3**

Current Technology Likes and Dislikes

| Likes | Dislikes |
|---|---|
| Speech capabilities | Too many levels of instruction in menus |
| Windows-like features and menus | Any features relying on sight |
| Shortcuts | Difficult keyboards |
| LookAround and localization features (similar to Sendero)[a] | Short battery life |

[a] Accessible GPS products by the Sendero Group; see http://www.senderogroup.com/products/GPS/allgps.htm.

of some technology applications was another prevalent barrier for adoption. One participant recalled situations where using guide dogs and technology aids at the same time made it difficult to navigate cluttered environments. Unfavorable weather conditions also hindered participants' ability to use mobile devices and navigation aids. However, in such situations mobile devices can be useful because they can be placed in a pocket and shielded from the weather. Travelers who are visually impaired need to pay attention to their surrounding environment and therefore require tools that do not interfere with that process. A good option to avoid blocking their hearing capabilities is to place electronic aids near just one of their ears. Bluetooth speakers were mentioned as a good alternative. Figure 7.4 summarizes the barriers to technology adoption that participants indicated.

### 7.5.7 Emergency Evacuation

Perhaps the most fear-provoking situation for a person with a visual impairment within an indoor environment is the case of an emergency requiring immediate evacuation. Available technologies do little to assuage this fear, and established emergency procedures typically do not account for people who cannot follow visual clues. Therefore, this is an area where technology innovation could significantly facilitate the safe passage of individuals who are visually impaired. In addition, tools created for this population could also apply to emergency rescue workers such as firefighters, who work in environments that hinder their vision.

#### 7.5.7.1 Preparedness and Prior Knowledge

In emergency situations, most respondents rely on previous knowledge to determine the closest safe exit. This knowledge may have come through personal inquiry, meetings, practice drills, or online materials. Many respondents take it upon themselves to find out about emergency procedures in new buildings, and one respondent stressed that this practice was out "of habit" and was not a result of vision impairment. In reference to unfamiliar

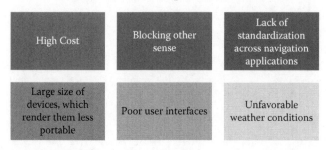

**FIGURE 7.4**
Barriers that preclude participants from adopting new assistive technologies.

environments such as a new hotel, some participants indicated that they would simply follow the crowd. Others reported that they would not know what to do in an emergency situation in an unfamiliar location. Thus, there is a need for additional assistance in unfamiliar situations in which individuals cannot rely on prior knowledge.

### 7.5.7.2 Technology Use

A very small minority of respondents reported using any type of technology during emergency evacuations. These tools include a smartphone app for tornado warnings and TTS used during emergency situations. The lack of assistive aids for emergency evacuation indicates a gap that can be filled by technology, particularly for unfamiliar locations such as hotels or large public buildings.

### 7.5.8 Perspectives of Building Managers

Most buildings have only front and rear entrances, so they mainly have a front desk attendant available to help visitors get to where they need to go, as well as keep unauthorized visitors out. The desk attendants typically warn visually impaired visitors during the check-in process about obstacles they might have a difficult time dealing with unless the visitor is escorted by someone else who works in that particular building. All of the buildings visited during this study had signage to help guide people to their destinations, emergency exit plans, emergency evacuation procedures, and braille translations underneath door number signs. However, none of these buildings had braille directional signs for the visually impaired.

Most of the building managers interviewed in this study had some experience interacting with individuals who are visually impaired, but they all agree that spontaneous events, unannounced constructions, and abnormal building structures can make it difficult for this population to navigate their buildings. Some of these abnormal building structures include half floors (e.g., half a dozen steps on the same floor), low walls, helix stairs, air bridges that connect two buildings, and events or activities on the premises. In addition, some organizations have a difficult time keeping their floor plans up to date because of a lack of funding.

Travelers who are visually impaired are usually expected to be able to navigate by themselves in most buildings. There are some indicators along the walls and high-contrast floors in some buildings, but it is difficult to measure how effective these methods really are to people with different degrees of visual impairment. Several building managers also appear to be afraid of offending people who are visually impaired by offering them assistance, so they often offer as little help as possible.

There is also a wide range of precaution and warning strategies for restricted areas between buildings. Some of the most common precaution methods are interacting with people (someone telling you verbally where not to go), discovering for yourself through signs, testing a door to simply see if it is locked, or just getting notices prior to entering the building. Most of the buildings visited during this study depend on security personnel to ensure that people without relevant authorization stay out of restricted areas; they do not use security cameras or other technologies for this purpose.

All of the managers we interviewed believe that technology applications will be useful in helping them to maintain accurate floor plans and to make these floor plans accessible to visually impaired visitors. Some also mentioned including a multi-extension file importer tool to enable adding different types of drawing files and voice navigation instructions that can further assist travelers who are visually impaired as they navigate these buildings.

## 7.6 Conclusions and Future Research

Travelers who are visually impaired have several concerns when navigating unfamiliar environments. First, they preplan their navigation routes as much as possible and need to build a mental map of the new environment they will be navigating. Once in a given indoor space, visually impaired individuals need to orient themselves in that location so they can identify where they are at any given time. Next, they need to figure out how to navigate to and from the location(s) of interest from a known environment. They also need to be informed of dynamic changes to the unfamiliar environment that may impact their safe and efficient wayfinding and navigation. Furthermore, they need to be able to "record" their navigation experience for future trips and also potentially share this information with others who might find it useful. Finally, if they get into any unsafe or difficult situation while navigating the unfamiliar environment, they need to have a reliable means of getting help.

Findings from our needs assessment support conclusions from previous work in this area and also align with the methodology adopted by O&M experts who work with visually impaired individuals. To orient themselves in an indoor environment and help identify their location, visually impaired people rely heavily on landmarks and clues around them. These include elements such as doorways, changes in elevation, and distinct sounds and smells. During navigation, travelers who are visually impaired depend on a mental or cognitive map of the space that they may have previously constructed and also use methods such as counting the number of landmarks or clues in their environment. If lost or disoriented, they typically need to ask for assistance from people nearby, although instructions from the average sighted person are usually not specific or detailed enough to be useful. Wayfinding maps can be extremely useful to visually impaired individuals, but they need to be accurate and up to date in order to be effective. Services such as ClickAndGo offer reliable and high-quality map information for visually impaired people. However, since this tool requires the map creator to personally visit each site and manually encode the building layout, it is not easily scalable to a larger number of indoor locations of interest. Building managers included in this study further highlighted some of the shortcomings of indoor spaces in terms of accessibility for visually impaired individuals. These included insufficient braille signage to provide directions to travelers and lack of procedures to prepare visitors who are visually impaired for dynamic changes in their environment. Above all these challenges, however, is the danger associated with emergency situations requiring evacuation from indoor environments. In such situations there are very few recourses available to visually impaired individuals who are not typically accounted for in standard emergency procedures and often have no guidance on how to exit a building safely and quickly in an emergency. This is perhaps the most critical need highlighted by this population that is currently unmet by any available techniques or tools.

Most of the participants included in our study were comfortable with utilizing technology to aid in their various activities, including navigation. Mobile phone technology was the most widely used tool, but several other devices and applications (such as GPS and PDAs) also play a role in the lives of these travelers. When working with new technology, most participants required some training, but the level and amount of guidance needed varied depending on the individual's skill and experience. In terms of technology preferences, audio input and output modes were clearly the most popular among participants. However, a potential challenge with audio output is that it could interfere with visually impaired travelers' ability to detect other auditory clues in their environment. Some participants also identified button and touch screen input modes and vibration or a combination of vibration and audio output modes as preferences. Customizability was a salient request among participants in terms of features they would want included in assistive navigation technology. More specifically, visually impaired travelers would like to be able to choose the level of detail included in directions and frequency with which instructions are provided. Although technology has the potential to address many navigation needs of visually impaired individuals, there remain barriers to greater levels of technology adoption among this population, including high cost, limited portability, output that blocks other senses, poorly designed user interfaces, and lack of standardization across tools. These challenges will need to be accounted for in designs of future assistive navigation technology.

While this study captures a detailed account of the indoor navigation needs of the visually impaired, it is limited in geographical scope to the Pittsburgh city of the United States. Although we do not anticipate drastic differences in navigation needs of these travelers in other parts of the country, there may be additional needs and concerns uncovered through similar studies across the United States. Therefore, potential future work includes extending this study to include a broader geographical sample of participants and collecting a sufficient volume of data to enable statistical analysis alongside a rigorous qualitative assessment. Beyond this enhancement, researching indoor navigation and wayfinding challenges faced by individuals who are visually impaired in other parts of the world will also add significant insight to this area of research, especially since many people now have the option to travel internationally. In general, gathering information from a pool of participants that is more representative of this population as a whole would better inform future developments of effective assistive navigation techniques and tools for this group of travelers.

Ultimately, indoor wayfinding and navigation tools should also seamlessly integrate with outdoor navigation guidance, transit assistance, and other assistive services that enable visually impaired travelers to safely and independently traverse urban environments. This means that urban navigation aids must incorporate accessible interfaces that allow people with disabilities to both receive and convey information and must be customizable to accommodate individual preferences. These tools must also be capable of indoor and outdoor localization at the resolution necessary for visually impaired travelers. Access to maps and other information in a variety of forms will also be critical so that routes that adhere to sensory and ambulatory constraints can be planned accordingly. Finally, to truly empower travelers who are visually impaired, these aids should provide mechanisms for advocacy to improve accessibility within the larger framework of the city infrastructure.

## Acknowledgments

Many individuals and organizations supported the work presented in this chapter. Preparation of this chapter was supported by the US Department of Transportation University Transportation Centers Program, Google Inc., and Carnegie Mellon University's Berkman Faculty Development Fund. Two of the authors were partially supported by the National Science Foundation under NSF-NRI Award Number 1317989. Any opinions, findings, and conclusions or recommendations expressed in this material are those of the authors and do not necessarily reflect the views of any sponsor.

Partner involvement has been essential for this work, and we were fortunate to collaborate with several groups that have supported various aspects of this project. Colleagues from the Blind & Vision Rehabilitation Services of Pittsburgh (BVRSP) and the Western Pennsylvania School for Blind Children (WPSBC) have been involved in needs assessment and encouraged their networks to participate in our research. We also wish to thank colleagues at the Western Pennsylvania School for the Deaf (WPSD) and the Helen Keller

National Center for Deaf-Blind Youths and Adults for participating in the early stages of this work.

Many individuals participated in our needs assessment, generously volunteering their time. Because of research compliance procedures, we cannot name these participants individually, but we are grateful for their involvement and enthusiasm for the project.

The authors also thank the many members of the research team who assisted with needs assessment, notably, Dr. Balajee Kannan, M. Freddie Dias, Dr. Yonina Cooper, Anna Kasunic, Dr. Gary Giger, Sam Jian Yu Li, Hannah Flaherty, and Soyeon Hwang.

Although authors Hend K. Gedawy and Sarah M. Belousov are now employed elsewhere, their contributions to the work presented in this chapter occurred during their time at Carnegie Mellon University.

## References

Golledge, R. G., Marston, J. R., Loomis, J. M., & Klatzky, R. L. (2004). Stated preferences for components of a personal guidance system for nonvisual navigation. *Journal of Visual Impairment and Blindness*, 98(3), 135–147.

Karimi, H. A., Nwan, D., & Zimmerman, B. (2009). Navigation assistance through models or experiences? *GIM International*, 23(12).

Long, R. G., & Giudice, N. A. (2010). Establishing and maintaining orientation for mobility. In W. R. Wiener, R. L. Welsh, & B. B. Blasch (Eds.), *Foundations of orientation and mobility* (3rd ed., Vol. 1, pp. 45–62). New York: AFB Press.

Orlando, A. C. (2009a). Angie C. Orlando—Deaf-blind and determined: The nose saves the day [Web blog post]. Retrieved April 11, 2011, from http://dotbug3. blogspot.com/2009/02/nose-saves-day.html

Orlando, A. C. (2009b). Angie C. Orlando—Deaf-blind and determined: Transporation again [Web blog post]. Retrieved April 11, 2011, from http://dotbug3.blogspot. com/2009/05/transporation-againg.html

Orlando, A. C. (2010a). Angie C. Orlando—Deaf-blind and determined: Battle zone [Web blog post]. Retrieved April 11, 2011, from http://dotbug3.blogspot. com/2010/05/battle-zone.html

Orlando, A. C. (2010b). Angie C. Orlando—Deaf-blind and determined: Sensory appreciation [Web blog post]. Retrieved April 11, 2011, from http://dotbug3. blogspot.com/2010/05/sensory-appreciation.html

Quiñones, P.-A., Greene, T., Yang, R., & Newman, M. (2011). Supporting visually impaired navigation: A needs-finding study. In *CHI '11 extended abstracts on human factors in computing systems* (pp. 1645–1650). New York, NY: ACM.

Sánchez, J., & Elías, M. (2007). Guidelines for designing mobility and orientation software for blind children. In *Proceedings of the 11th IFIP TC 13 international conference on human-computer interaction* (pp. 375–388). Berlin, Heidelberg: Springer-Verlag.

Williams, M. A., Hurst, A., & Kane, S. K. (2013). "Pray before you step out": Describing personal and situational blind navigation behaviors. In *Proceedings of the 15th international ACM SIGACCESS conference on computers and accessibility* (pp. 28:1–28:8). New York, NY: ACM. doi:10.1145/2513383.2513449

Zimmerman, G. J. (1990). Effects of microcomputer and tactile aid simulations on the spatial ability of blind individuals. *Journal of Visual Impairment and Blindness, 84*(10), 541–547.

# 8

# Indoor Navigation Aids for Blind and Visually Impaired People

**M. Bernardine Dias**

**Satish Ravishankar**

## CONTENTS

*Abstract:* The challenge of enabling safer and more effective indoor navigation for visually impaired people has motivated researchers for decades. However, overall, there is a lack of practical, low-cost, accessible, customizable, and user-friendly navigational aids to allow people with low or no vision to safely navigate or evacuate buildings. The existing solutions (both commercial and research) have been developed to assist blind and visually impaired people to familiarize themselves with different physical spaces, localize themselves within a space, plan useful routes within a space, and avoid hazards while moving from place to place. Among these assistive tools are enhanced mobility aids, which include tools such as augmented white canes with laser-emitting diodes and sensors, magnetic field probes, radio-frequency identification, ultrasonic devices, and camera and vision-based systems. Location-aware systems have also been developed using technologies such as GPS, mobile phone infrastructure, or wireless access points to triangulate the location of a user. Some of these technologies for localization have been coupled with tracking tools to create navigation systems for visually impaired users.

However, existing assistive indoor navigation tools have a few critical shortcomings. First, there are currently no available low-cost localization solutions that can provide the resolution and accuracy required for the task. Second, most of the navigation technologies do not provide routing instructions based on accessible landmarks in the navigation environment. Finally, flexibility and customization of features based on user preference is not explored in depth, despite being a critical need for people with visual impairments. In this chapter we survey the state of the art in assistive technology solutions for enabling visually impaired people to navigate unfamiliar indoor environments. We also identify the gaps in the available technology and explore the technical challenges in closing these gaps.

## 8.1 Introduction

The most ubiquitous and universally identifiable navigation aid for blind or visually impaired (B/VI) people is the white cane. B/VI people have been using some form of cane to guide them and warn others for as long as we know (Kelley, 2009). Even the earliest publications analyzing mobility needs of the B/VI community include guidelines on how best to use a cane (Kelley, 2009). However, the now-ubiquitous long cane and its current use were not developed until post-World War II, when the profession of orientation and mobility was founded to treat those blinded during the war (Kelley, 2009).

The use of guide dogs (currently referred to as guide dogs) also began as a postwar measure. Germany pioneered this practice by training German shepherds to guide those blinded during World War I (Kelley, 2009). This mode of travel for B/VI people was brought to the United States in the early 1920s and became an established procedure in the United States in the late 1920s once the Seeing Eye (now the most prominent and well-regarded dog guide training institution in the country) was established (Kelley, 2009).

Technologists also joined the effort to improve the lives of the B/VI population from early stages with initial work in creating electronic reading and writing aids. The 1950s show several patents on electronic mobility aids for B/VI people, and in 1964 Kay published one of the earliest research papers on the topic, describing an electronic navigational aid in the form of an ultrasonic sensing probe. Many related research efforts followed, and today assistive technology is a key topic in several international technology conferences and journals.

While assistive technology for enhancing navigation capabilities of B/VI people has been a popular research topic for decades and has yielded many useful outcomes, the number of practical ubiquitous tools produced has been low because of numerous factors, including the wide range of requirements among this user population. Among the tools that are currently in use or show strong potential for this path, few are useful in the context of navigating indoors or in GPS-denied environments. Nevertheless, indoor wayfinding and navigation remain a significant challenge for a large component of the B/VI population, especially in the urban context. Therefore, the quest for assistive technology that can enhance indoor navigation capabilities of the B/VI population remains an important research endeavor. This chapter surveys the work to date in assistive indoor navigation technology solutions for B/VI people and identifies the remaining gaps in this area of research. The survey does not aim to be comprehensive in coverage of all relevant research publications but instead provides an overview of the key categories of technology and tools being researched and developed in this field, along with examples from the literature in each of these categories.

## 8.2 Indoor Navigation Technology Challenges

An indoor navigational aid for the blind generally has to perform one or more of the functions shown in Figure 8.1: familiarization, localization, path or route planning, and communicating with the user in a meaningful manner through an accessible interface. Each of these functions is explored in greater detail in the following sections.

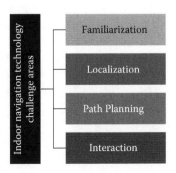

**FIGURE 8.1**
Technology challenge areas in indoor navigation for B/VI people.

## 8.2.1 Familiarization

The ability to safely and independently explore a new environment goes a long way toward improving a person's quality of life. When exploring unfamiliar environments, sighted people mainly use visual information to create a mental map. Some sighted individuals may not consciously make the mental map (or may not be aware of the fact that they are making a map) but still absorb "visual" cues from the environment and memorize those cues that help them get familiarized with the location. Without the use of visual information, exploring unfamiliar environments can sometimes become a hazardous task for the blind. Many visually impaired people are reluctant to explore new places, especially if they have not gained sufficient self-confidence or expertise navigating within that type of environment. Therefore, familiarization with an environment is a key factor in enhancing the safety and independence of B/VI people during indoor navigation. While B/VI people gain this familiarization by using the guidance of an orientation and mobility instructor, using tactile maps (Espinosa, Ungar, Ochaíta, Blades, & Spencer, 1998), or relying on the help of sighted friends, technology can also play a role in increasing the opportunities for B/VI people to familiarize themselves with indoor spaces.

B/VI people who cannot use information from visual stimuli use information from other sensory channels such as tactile, olfactory, and auditory stimuli. The auditory information from their surroundings can be used to detect activities and people by voice or, less frequently, for echolocation.[*] The sense of smell can, in some situations, provide useful information about the current location, such as the unique smells of a bakery. Tactile information is gathered using a cane, hands, and feet (to detect the texture of the floor) and forms a major portion of the mental map. While it is not a substitute for experiencing the real space with the guidance of an orientation and mobility instructor, a well-designed virtual navigation tool can allow B/VI people to remotely explore an unfamiliar environment and build an initial cognitive map of the space (Lohmann, Yu, Kerzel, Wang, & Habel, 2014). This could lead to greater independence because of their virtually acquired familiarity with the environment and faster acclimatization, which often leads to safer exploration. Providing all of the needed cues in a scalable and sustainable manner through a virtual environment is, however, not an easy task. In addition, because of the wide range of landmarks and clues that can be used, and because of the wide range of visual impairments and preferences for different forms of guidance in the B/VI community, creating a virtual environment that accommodates all of these constraints is a significant challenge.

---

[*] Some humans can detect objects around them by sensing echoes from those objects. This technique is used by some blind people for navigating, similar to animal echolocation used by bats and dolphins. Blind people who use echolocation typically create sounds by tapping their canes or snapping their fingers and then interpret the sound waves reflected by nearby objects to identify their location and size.

## 8.2.2 Localization

In the context of indoor navigation, a localization system assists a user in identifying his or her location (and orientation in some cases) within the indoor environment. Various methods are used for localization indoors, but GPS does not form a major part of these methods because walls and roofs usually occlude line-of-sight communication with satellites, and these obstacles attenuate the signal as well. This makes indoor localization more of a challenge compared with outdoor localization.

The traditional approaches that do not employ technology are the use of a mental map built through guided exposure to the environment or through auditory instructions, and the use of tactile maps (Subryan, 2009). Maps can be advantageous in their flexibility of size while providing a visually impaired traveler with a comprehensive representation of an environment, catered specifically to the needs and constraints of that user. However, these maps lack the ability to dynamically provide the user with feedback during navigation and cannot be easily customized for people with a variety of visual impairments or updated to reflect current information. Large physical maps can also be cumbersome to be carried around during navigation and hence are rarely used in a portable manner.

A variety of other techniques are being explored to achieve indoor localization, some of which require alterations to the indoor environment or infrastructure that are not ubiquitous. Torres-Solis, Falk, and Chau (2010) reviewed a variety of indoor localization technologies, as follows:

- *Radio frequency:*
  - *Personal and local area networks:* Wi-Fi signals
  - *Broadcast and wide area networks:* mobile phone signal
  - *RFID tags*
- *Photonic:*
  - *Visible light:* visible light communication and computer-vision-based methods
  - *Infrared*
- *Sonic waves*
- *Mechanical*
  - *Inertial:* using a sensor like a gyrometer, an accelerometer, or an inertial measurement unit
  - *Atmospheric pressure*
- *Magnetic:* using sensors like a compass to determine heading

Several approaches use vision- or camera-based systems ("Project Tango," n.d.; Ravi, Shankar, Frankel, & Elgammal, 2005). Typically, these approaches use a database of images of the indoor environment or other visual tracking algorithms to perform localization. There are also systems that use RFID tags, visible light communication, infrared or ultraviolet signals, and Bluetooth beacons (Ganz et al., 2012; Giudice & Legge, 2008). This requires installation of new infrastructure in a building. For example, RFID tags may need to be installed throughout the indoor space that needs localization features, and these serve as landmarks for localization. Similarly for visible light, infrared, or ultraviolet signal-based localization, LEDs, infrared, and ultraviolet signal emitters, respectively, have to be installed in the buildings of interest.

Indoor localization techniques can be broadly categorized into those that require installation of new infrastructure (infrastructural) and those that do not need any such installation and use preexisting infrastructure. The latter category of approaches, in general, use one or more of the following techniques for indoor localization: dead reckoning, GPS, visual simultaneous localization and mapping (SLAM), and Wi-Fi RSSI (received signal strength indication).

GPS (in the limited cases where it can be detected indoors) and Wi-Fi RSSI data within the building provide indoor location landmarks. In the case of Wi-Fi RSSI, a database of Wi-Fi RSSI data is required for points on an indoor map. These points are given a Wi-Fi fingerprint using the Wi-Fi signals and their corresponding strengths. As the user travels through an indoor space, any time there is a fingerprint match, the user can be localized using the database of Wi-Fi fingerprints. Wi-Fi RSSI data can therefore be used for triangulation or fingerprinting (Quan, Navarro, & Peuker, 2010). The fingerprinting method involves using a probability distribution of signal strengths on a map at every location. Given a set of sampled signal strengths, the probability distribution map can be used to predict the corresponding location. Triangulation is a simpler method that relates signal strength as a function of distance from a Wi-Fi access point to perform localization. A major problem with using Wi-Fi RSSI data is the inherent nature of these signals to fluctuate with time. Moreover, with the increased use of access points that adapt their signal strength according to the current load and interference caused by mobile phone Wi-Fi radios, the challenge of using this technique for localization is made more difficult. To reduce the fluctuations, Wi-Fi fingerprinting usually considers the dominant access points in an environment rather than all of the visible access points. A system that utilizes GPS exploits recorded GPS-based location before the person enters a building or occasional GPS hits within the building for use as landmarks. As the user travels between these landmarks, dead reckoning (which comes under inertial measurements) is used to keep track of distance and direction of travel.

SLAM is a technique used in robotics to enable robots to explore an environment, build a map, and localize itself in the map (Karlsson et al., 2005). The technique of visual SLAM uses cameras to acquire data from the environment and then utilizes a combination of computer vision and odometry algorithms to map the surrounding space, which enables robots to autonomously explore their environment (Karlsson et al., 2005). Smartphone cameras are becoming increasingly powerful and affordable, and smartphones are simultaneously incorporating high-performance computers that have the necessary computing power to effectively use visual SLAM techniques (Karlsson et al., 2005). This trend is a strong indicator that visual SLAM will be one of the main contributors to better localization systems in the near future.

### 8.2.3 Path Planning

In addition to localization, an indoor navigational aid must be capable of path planning and communicating the path to the user. Localizing the user and planning the path to the user's desired destination go hand in hand. Once a user has been localized, the optimal path to the destination can be determined and communicated to the user as accessible instructions. There is always a possibility that the user may veer from the recommended path for many reasons, and a smart navigation aid will be capable of dynamically replanning the path to the user's destination based on his or her new location.

Most navigation systems for sighted users choose the shortest path. However, this may not be ideal when the user is visually impaired. For example, the shortest path may have higher probability of a B/VI user veering away from the planned path compared to a path that goes along walls that can be used to maintain direction and orientation (Fallah, Apostolopoulos, Bekris, & Folmer, 2013). The directions must include landmarks that can be sensed during navigation by the B/VI user while remaining simple and effective. The navigation system should also take into consideration all the environmental information used by the visually impaired to orient themselves. An example of the kind of instructions given by orientation and mobility instructors to visually impaired travelers is shown in Figure 8.2.

Furthermore, any change in an indoor environment may confuse a B/VI user since some of the landmarks and clues they use for wayfinding may have been altered or lost. For example, the smell of coffee at a particular

---

The lobby doorway consisting of a set of outer and inner automatic doors separated by a narrow vestibule is located in the center of the east wall of the building. Upon entering through the inner doors, walk forward 6 feet over a carpeted area. Once the floor surface changes from carpeting to tile, turn left and walk forward 28 feet. This is the southeastern corner of the lobby. Turn to the right and walk forward 25 feet. The center of the front desk will be directly to the left. Turn to the left to face the desk. A small area directly in front of the desk is carpeted. Note that three large square supporting pillars bisect the walkway that runs along the southern end of the lobby, so it is best to keep to the outer edge when traversing the lobby; that is, the left when approaching the desk from the southeast corner or to the right when crossing the southern end of the lobby from west to east.

---

**FIGURE 8.2**
An excerpt from directions for B/VI people to a front desk in a hotel lobby provided by orientation and mobility instructor Dr. George J. Zimmerman from the Vision Studies Program, University of Pittsburgh, Pennsylvania, United States.

place in a building may be a clue that is used by a B/VI navigator to identify the location of a coffee shop. If that coffee shop is relocated or converted into a stationery store, it may leave the B/VI traveler confused. Although this environmental change may not pose a problem to the navigation system, this becomes a relevant change to communicate to the B/VI user of the navigation system. Therefore, navigation systems for B/VI users must be able to incorporate accessible environmental landmarks and clues into their instruction sets and notify users of relevant changes to the environment as needed.

### 8.2.4 Communication and Interaction

Once an appropriate path to the destination has been planned, the navigational aid should translate the path into directions that a B/VI user can follow and communicate these directions to the user in an accessible and nonintrusive manner. The translation and communication have to be customizable to the constraints and needs of the B/VI user. Moreover, the method of communicating these instructions to the user should not distract the user from paying attention to environmental landmarks and clues that he or she uses to navigate.

The preferred method of receiving instructions from a navigation aid for most B/VI users is audio or a combination of audio and vibrations.* As for providing the system with input, the preferred method is again audio using voice instructions. B/VI users are also typically willing to interact with technology through touch-based input methods that use tactile buttons and/or relevant accessible techniques for the use of touch screens. Although current technology allows such methods of communication with a variety of portable devices, there are still a few issues that remain unresolved. For example, using voice-based input methods in a crowded area is still not reliable since the system may not be able to recognize the commands because of high levels of background noise. Some B/VI people are also concerned about using touch-screen gestures because of the worry that they will draw the gestures incorrectly. Another concern with receiving audio instructions from the device is whether these instructions will interfere with the user's interaction with the environment and his or her ability to pay attention to environmental cues. To allay some of these concerns, assistive navigation devices must allow the users to listen to the device through only one ear and easily vary the verbosity of the instructions based on the situation. In cases where a combination of audio and vibrations are used in the communication interface, the device should provide the user with an option to switch between using only vibrations, using only audio, and using a combination of the two interaction methods.

---

* See Chapter 7 in this book for further details on the information provided in this paragraph.

## 8.3 Example Indoor Navigation Aids

The previous section in this chapter examined the key challenges technology tools must address to enable safe and independent indoor navigation for B/VI travelers. This section reviews several tools used as indoor navigation aids for B/VI people. The reviewed categories of technology aids are illustrated in Figure 8.3.

The first subsection reviews familiarization tools, which are tools or methods that provide users with information about an environment before they physically explore it. The following subsections review tools developed for use during navigation in the categories of (1) enhancements of traditional mobility aids, (2) smartphone-based solutions, and (3) custom-designed navigational aids for the blind. The final subsection reviews infrastructural enhancements required by some of these solutions.

### 8.3.1 Familiarization Tools

Technologies such as virtual navigation environments and narrated maps have demonstrated great potential to encourage and assist B/VI people with familiarization, safe exploration, and navigation of indoor environments. Narrative maps (Sheepy & Salenikovich, 2013) are one approach to familiarizing B/VI people with an indoor environment prior to physical interaction

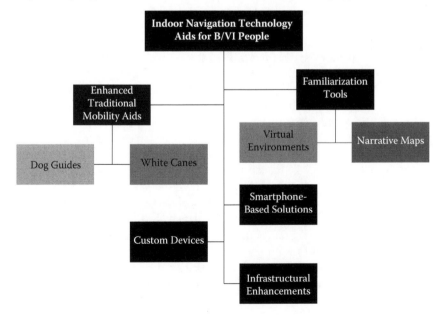

**FIGURE 8.3**
Indoor navigation technology aids for B/VI people reviewed in this chapter.

with that space. To create a narrative map, an orientation and mobility expert would normally describe the indoor environment and carefully include any sensory landmarks or clues that will be useful for wayfinding. This usually includes an overall description of the environment, followed by a description of the paths to be taken to various locations within the environment. This helps the B/VI traveler form an initial mental map of the environment based on this narrative map.

While several technologies use auditory descriptions and have attempted to automate the creation of narrative maps, no such solution is ubiquitous because of the many challenges entailed in extracting the relevant information about an environment and presenting this information in accessible form to B/VI users. "Directions" (Sheepy & Salenikovich, 2013), a smartphone application, is one such attempt. This navigational aid allows blind users to use a series of prompts through an accessible touch screen interface to get directional guidance and instructions from a sighted user. ClickAndGo Wayfinding Maps* eliminates the need for real-time help from a sighted user. It, however, requires any location (indoor or outdoor) to be manually surveyed before this service can be provided. After extensive surveying, navigation instructions are prepared and recorded. A visually impaired person who wishes to go to an area of interest in a location can then simply enter start and destination landmarks through a portal on the website and gain access to detailed instructions that can be downloaded in text or audio format. An example of a completely automated approach to providing narrated maps is StreetTalk GPS.† Users are allowed to search for a route to a destination from their current location or form a location of interest. StreetTalk then plans the route and provides turn-by-turn instructions that are announced using voice-based commands and/or braille. It also provides a virtual navigation mode in which the user is allowed to explore the map or a certain route as though he or she were a pedestrian. Trekker (National Federation of the Blind, n.d.; "Trekker Talking GPS," n.d.) is yet another GPS-based navigation aid for the blind that provides automated speech-based detailed directional instructions that include information about cross streets and even informs the user if a street is two-way or not. Similar to StreetTalk, it also provides a virtual exploration mode that can be used either online or offline to traverse through locations of interest using the arrow keys on the keyboard. Trekker and StreetTalk are both designed for outdoor navigation. There is a need for an equivalent tool that provides similar services for indoor locations.

Narrative descriptions alone can be insufficient to familiarize B/VI travelers with unfamiliar indoor environments. A richer solution is to provide B/VI users with the ability to virtually navigate a digital representation of the environment. Using such a virtual exploration tool, B/VI users can build a detailed cognitive map of an environment before physically interacting with

---

* See http://www.clickandgomaps.com/narrative-maps.
† See http://www.freedomscientific.com/products/fs/streettalk-gps-product-page.asp.

the environment. This could lead to a greater degree of independence because of their virtually acquired familiarity with the environment and faster acclimatization. The virtual environment could be represented using sounds, as in the work by Sánchez, Sáenz, Pascual-Leone, and Merabet (2010). The authors represented an environment using an Audio-Based Environments Simulator (AbES) and studied the changes in the brain while blind participants simultaneously explored the environment. With AbES, every object in the virtual environment has an associated sound in addition to spoken audio cues. From their studies, the authors concluded that the use of virtual environments can be highly effective for the purposes of training and rehabilitating the visually impaired in the domain of navigation. Picinali, Afonso, Denis, and Katz (2014) conducted a virtual navigation study in which they represented virtual architectural spaces using audio events and allowed blind participants to interact with this virtual environment. They discovered that the mental map of a virtually navigated space preserved topological and metric properties as is expected in the case of actual physical navigation.

Lahav and Mioduser (2004, 2008) enhanced the virtual environment concept further and created a multisensory experience for blind users to explore. A force-feedback joystick was used to guide the B/VI users as they explored the virtual environment. The virtual environment was based on a physical environment and consisted of objects modeled using the objects in the actual environment. These objects were associated with vibration, attraction, or repulsion effects when the users interacted with them in the virtual world. The virtual environment also associated objects with auditory labels such as birds chirping to indicate a window and explicit names of objects (e.g., pillar 1 or box 3) to help the B/VI users create a cognitive map more easily. When this environment was tested with blind participants, they were allowed exploration without any time constraints. The virtual exploration of the environment was then compared to physical exploration done by control participants. The researchers found that participants engaged in virtual exploration of the environment for approximately four times the duration and traversed approximately three times the distance in comparison to the participants in the control group. This study indicated that although participants from the control group were better able to verbally reconstruct landmarks, the experimental group did better in reconstruction accompanied by virtual drawings of the location of the landmarks. One can conclude that the experimental group developed a better perception of the space, position, and location of objects that they interacted with. Overall, the study concluded that virtual exploration led to participants exploring the physical environment in a relaxed and safe manner.

With further work to represent complex public spaces, irregular surfaces, indoor and outdoor spaces, and a variety of landmarks and clues commonly used by B/VI people, virtual representation of environments can be improved to better suit the needs of the B/VI population and to reduce the time they require to familiarize themselves with unfamiliar environments.

## 8.3.2 Enhanced Traditional Mobility Aids

The long white cane and dog guide are still the most ubiquitous assistive tools used by the B/VI population, and hence many attempts have been made to augment these traditional mobility aids with technology. Wearable devices based on modifications of sunglasses have also been developed. There are a number of modified versions of the cane currently on the market that provide added sensing capabilities to the user. Some of the modified versions employ sensors to detect obstacles, and this information is relayed to the user via tactile or audio feedback. One product in particular, the Sensible Blind Cane (Centre for Innovation and Development, n.d.), employs an ultrasonic sensor module to detect obstacles and is designed to be low cost, which increases affordability. Another instance of the modification of the simple cane is the MobiFree set of tools (Lopes, Vieira, Lopes, Rosa, & Dias, 2012). This set of tools includes the MobiFree Cane, MobiFree Sunglasses, and MobiFree Echo. The cane uses ultrasonic sensors to detect obstacles and irregularities on the ground. When an obstacle is detected, the user is warned using vibratory tactile feedback. It also has the added functionality of being able to detect ambient light conditions and switch on a set of LEDs, when needed, to make the blind user more visible to any sighted people in the surrounding area. The sunglasses use the same technology as the cane but detect obstacles at eye level. SonicGuide (Kay, 2000) is another example of enhanced sunglasses that use wide-band ultrasonic waves to get spatial information of the scene in front of the user and communicates this as audio information in the form of complex tones. It allows the users to receive information about their environment, enabling them to distinguish between scenes and perceive obstacles. The most recent model of the SonicGuide is KASPA.

Guide dogs are yet another example of a traditional navigation aid for B/VI people. Even though guide dogs provide users with a great degree of mobility and freedom, this aid is restricted by the high cost of training the dogs and the need for very selective breeding (Galatas, McMurrough, Mariottini, & Makedon, 2011; Xu & He, 2012). This has led to research that attempts to replace the guide dog with a robot. For example, Xu and He (2012) developed an ARM-processor-based intelligent guide vehicle equipped with ultrasonic and photoelectric sensors. The ultrasonic sensors detect obstacles on the road, and users are notified about their distance from the obstacle through a voice prompt. If the obstacle is closer than a set threshold (50 cm or ~19.68 inches in this case), the system has the capability to automatically guide the user safely around it. The photoelectric sensor is used for the purpose of road detection and tracking in order to prevent the user from straying from the path. eyeDog (Galatas et al., 2011) is an approach to replace guide dogs with an assistive-guide robot that uses vision-based navigation systems and laser-based obstacle avoidance techniques. The system is composed of a netbook, a USB webcam, and a LIDAR (light detection and ranging) unit. The images captured by the camera are used to ensure that the user is guided by the robot in a direction parallel to the corridor or road, and the LIDAR unit is used to detect and avoid obstacles.

Despite the many attempts to enhance these traditional navigational aids, the traditional white cane and the guide dog remain the most ubiquitous navigation aids used by B/VI people. Therefore, the technology enhancements to these aids need to be either improved in a variety of ways or better distributed to increase their ubiquity and impact.

### 8.3.3 Smartphone-Based Mobility Aids

Smartphone-based outdoor navigational aids for the blind that use GPS have been developed by several groups and are currently used by many B/VI people. BlindSquare[*] is one such application developed for iOS devices that makes use of data from Foursquare and Open Street Maps to help users locate stores and cafés around them. Loadstone GPS, Mobile Geo, and Seeing Eye GPS are further examples of GPS-based systems that function as navigational aids for the visually impaired in outdoor environments.[†] However, smartphone-based indoor navigation aids are more challenging to develop. This section describes some of the research efforts that attempt to achieve this.

Some indoor navigation aids need a database of building signatures defined by sensor readings or a combination of sensor readings correlated with location, while others need no predeployment effort such as setting up additional infrastructure or creating a database. For example, Wang et al. (2012) developed a system that requires predeployment. It uses all of the sensors (gyroscope, accelerometer, etc.) in a smartphone to characterize a building by different signatures in different locations. These signatures are used as landmarks to determine the location of a device or user. Between landmarks, dead reckoning is used, and these location signatures are then used to correct the errors accumulated in dead reckoning. In this approach, problems such as electromagnetic variations in a specific part of a building (which usually affects specific sensor readings) are used as part of the signatures. A database of these signatures is required before this method can be deployed, but it is not clear how often this database needs to be updated. Furthermore, because this work uses a variety of sensors on the smartphone to detect location signatures, the database may store signatures that require sensors that are not available on some smartphones.

To solve the problem of indoor localization, Ravi et al. (2005) used visual tags, which also require an extensive predeployment effort. Images are captured by the user's smartphone and periodically sent to a server. The server localizes the user by comparing these images with those already in its database. This method therefore requires extensive image collection throughout the target

---

[*] See http://blindsquare.com/about/.

[†] See http://www.loadstone-gps.com/; http://www.codefactory.es/en/products.asp?id=336; https://www.senderogroup.com/products/shopseeingeyegps.htm, respectively.

indoor space, along with a potentially large database that will have to be periodically updated. Also, since this method involves image processing, it could require more processing time and computing resources compared with other methods. Another recent example of an approach that utilizes the smartphone camera is Audible Vision developed by Yu and Ganz (2012). It uses computer vision to estimate the user's position and orientation and then provides the user with information about the location of obstacles in the environment relative to the user in a 3D audio interface. This system requires the predeployment effort of creating a database of photos that completely represent the environment. From an initial set of images, the environment is reconstructed, and a subset of images that cover all extracted features is used during the localization phase. This system has been implemented on the Android platform and is capable of performing sensor management (to determine camera rotation), localization using these images, and generation of 3D sound to identify the obstacle locations. In this method of feedback, the audio description of the location, characteristics, and distance of obstacles virtually appears to come from the location of the obstacle itself. This system was designed not to replace a navigation system (since it does not provide navigation instructions) but to work in conjunction with one to improve performance.

Chintalapudi, Iyer, and Padmanabhan (2010) developed an indoor navigation system that does not require any predeployment effort. In this approach, users move around inside a building, and the phones transmit measured RSSI of Wi-Fi signals from access points back to a server. Occasionally, there will be a GPS hit near an exit or entry or a window along with the measured RSSI. This is also sent back to the server. This information is processed by a localization algorithm running on the server to accomplish localization. However, this approach depends on occasional, and sometimes improbable, GPS hits indoors, which will not work well in some locations (a basement for example).

Laoudias et al. (2012) used crowdsourcing to collect Wi-Fi RSSI data and neighborhood AP MAC addresses for indoor positioning. The participants recorded data by marking points on a map indicating their current location as they walked inside a building. The number of samples collected at each point on the map was preset by the participants before they started collecting data. The data were then added to a central database and used in a Wi-Fi fingerprinting algorithm to localize the user. If a large number of data points are collected, finer localization can be achieved, and this approach can be used for creating or enhancing a navigational aid for the blind.

Navatar (Fallah, Apostolopoulos, Bekris, & Folmer, 2012) is a cost-effective system designed for large-scale deployment. It attempts to provide navigational instructions to the user without augmenting a smartphone with external signal sources or other infrastructure. The system uses a virtual representation of the indoor environment that uses tactile landmarks (such as doors, walls, and hallway intersections) that the user can sense. Feedback from the user upon confirmation of landmarks in the environment is used as ground truth, allowing Navatar to periodically update location data. In between landmarks, dead reckoning (using smartphone sensors such as the accelerometer) is used to perform localization. The problem of accumulated error that comes with the use of dead reckoning is overcome by periodic inputs from the user whenever he or she detects a landmark.

These approaches all use ubiquitous smartphones to develop an indoor navigation aid for B/VI people. The use of a smartphone, which makes the system relatively affordable, comes with the restriction of having to use only the sensors available on commercial smartphones. This leads to problems with accuracy of sensor readings and limits the approach from utilizing technology not associated with smartphones. One way to overcome these issues is the use of custom devices designed specifically as navigation aids, which are discussed next.

### 8.3.4 Custom Devices

In addition to systems that use ubiquitous smartphones, several researchers have developed custom devices that B/VI users wear or carry as navigation aids. Drishti (Ran, Helal, & Moore, 2004) is an example of a custom device that uses a wearable computer designed to be a navigation aid for B/VI people. The system is designed to be a navigational aid both indoors and outdoors. In an outdoor setting, it uses differential GPS to localize the user and provides instructions that allow the user to travel safely on sidewalks. Indoors, it uses an ultrasound positioning system that provides an accuracy of 22 cm (approximately 8.6 inches). This system therefore requires additional infrastructure to be installed in indoor environments but is able to provide localization of sufficient accuracy that it can be reliably used as a navigational aid for the blind. The system is also capable of dynamic path planning and replanning.

PERCEPT, developed by a team of researchers at the University of Massachusetts (Ganz et al., 2012), is another system that requires additional infrastructure since it employs passive RFID tags embedded in the indoor environment to provide navigation instructions to blind travelers. When a visually impaired traveler, equipped with a smartphone and the PERCEPT glove, enters a building, he or she scans the destination location at a kiosk. The traveler is then guided to the chosen destination with navigation instructions using landmarks. The kiosks have raised letters indicating room numbers and location labels along with their braille equivalent. The PERCEPT glove, which has an RFID reader, Bluetooth radio, microcontroller, and related circuitry, allows the user to freely use his or her hand to read signs by touch while scanning RFID tags at the same time. The user has the choice of interacting with the PERCEPT system using buttons, the glove itself, or the phone. Navigation instructions are received by the phone over the server and relayed to the user after text-to-speech conversion. Along the same lines as the PERCEPT glove, the Wayfinding Electronic Bracelet (WEB) (Bhatlawande, Mahadevappa, & Mukhopadhyay, 2013) is a portable device that employs an ultrasonic transceiver mounted on a circular bracelet to perform object detection. The onboard processor runs a real-time system that provides the user with vibro-tactile and audio feedback about detected obstacles in the surrounding area through a motor and a buzzer, respectively.

Another example is the Digital Signage System (DSS) (Giudice & Legge, 2008), which employs a handheld device equipped with an infrared emitter that the user pans until a reflection is received from one of many retro-reflective bar codes strategically placed in the indoor environment. The bar code is read by the DSS using this reflection, and this information is fed to the building database (called the Building Navigator), which then returns to the user information about the content of the surroundings and routing to the destination using a synthetic voice as audio feedback.

The work by Hub, Diepstraten, and Ertl (2003) uses a portable computer, like Drishti, that is carried by the user. This system uses ultrasonic sensors and a stereo camera along with a 3D inclination sensor and a digital compass. The camera input is used to detect obstacles in the scene in front of the user and also gets information regarding object color, distance, and size, which can be used to suitably guide the B/VI user. PERSEUS (Personal Help for Blind Users) developed by Vítek, Klima, Husnik, and Spirk (2011) also uses a stereo camera and wearable computer and additionally incorporates input from a sighted individual. The visually impaired user wears protective acrylate glasses fitted with two cameras and an acoustic transducer. At times of distress, the user signals the navigation center, which then alerts a sighted operator. A stereoscopic video stream of the user's view transmitted to the navigation center via public Wi-Fi is used by the operator to guide the blind user by providing audio instructions.

Kaiser and Lawo (2012) designed a wearable navigation system that uses SLAM targeted at both indoor and outdoor environments. This custom device to be carried by the user has a short range laser, an inertial measurement unit, headphones, and a wearable computer. SLAM is used to build a map of the environment that is being explored and at the same time keeps track of the user's position in that environment. The system uses the constructed map to guide the user to the desired destination using audio instructions. The researchers also addressed the problem of conventional headphones blocking other auditory signals from the environment that the blind use to navigate. Instead of conventional headphones, audio bone headphones are used to unobtrusively provide navigation instructions. Audio bone headphones are bone-conduction-based headphones that do not block the ear canal. The sound is transmitted directly to the innermost part of the ear, bypassing most of the ear canal ("How Bone-Conducting Headphones Work," n.d.). This form of sound transmission leaves the ear canal unobstructed and capable of receiving auditory stimuli from the environment.

### 8.3.5 Supporting Infrastructure

As discussed previously in this chapter, infrastructural navigational aids may require installation of additional infrastructure in indoor environments to provide an increased accuracy of using solutions such as RFID tags, visible light communication, and so on. Some navigation aids described in previous sections of this chapter use infrared or ultraviolet signals from designated locations within an environment as ground truth to perform localization and error checking. For this purpose, LEDs (each with a unique signature) that emit visible, infrared, or ultraviolet lights need to be installed in the environment. An example of a system that uses infrared communication is DSS, described previously (Giudice & Legge, 2008).

Since smartphones are becoming increasingly popular as a platform for designing navigational aids, Bluetooth beacons and near field communication (NFC) tags, which are commonly used in conjunction with smartphones, can be termed as supporting infrastructure. These beacons and tags, when placed at strategic locations, can provide a mobile phone with location information and can supplement the use of Wi-Fi signals and onboard accelerometer and gyroscope (in smartphones) for navigation. NFC tags, which are based on RFID tags, have a significant advantage in that they do not require power for operation and hence will require much less maintenance.

With the active research area of Wi-Fi fingerprinting, Wi-Fi access points are a major part of the supporting infrastructure. Since most public buildings have Wi-Fi access, it is safe to say that access points need not be installed specifically for the purpose of creating navigational aids for the blind. But, as mentioned earlier, installation of adaptable access points and interference from Wi-Fi radios of mobile phones give rise to complications in using Wi-Fi RSSI signals for localization. This may lead to a situation where Wi-Fi fingerprinting can be used reliably only in conjunction with other technology such as RFID tags. Another supporting infrastructure that could be investigated is the use of visual tags physically placed inside buildings in strategic locations. These tags could be bar codes, signage for floors, hallway and room numbers, and room labels (conference hall, library, kitchen, etc.), all placed at a suitable height and location in buildings. These tags can be used for approaches that use smartphone cameras. Images captured through a smartphone camera as a user walks through a building can be analyzed using image processing (onboard the phone or offloaded to a server), and these tags could be detected and used to deliver location information to the user.

The infrastructure installations just described make use of technology that is cost-effective and show great promise in making indoor spaces accessible to B/VI people. RFID and NFC tags and LEDs installed in a building could help not only visually impaired travelers but also the general public if they can be exploited for use by location-based services for the general population. For example, RFID tags installed in a grocery store can be used as landmarks indicating different aisles, or they could be used to provide shoppers with information about the contents of the aisle and special offers available in that aisle.

In time, some of these infrastructural enhancements may be incorporated into accessibility standards mandated for indoor environments. For example, accessibility standards for buildings that require installation of NFC or RFID tags according to a strategic placement plan could be introduced. Since such infrastructure is generally flexible in the ways it can be exploited for use by sighted people as well, it may not be long before these infrastructural standards are adopted more widely for implementation in indoor spaces.

## 8.4 Conclusions

All research efforts described in this chapter are motivated by the desire to improve the quality of life for the visually impaired. However, the accessibility and viability of these solutions remain to be proven. Some of the methods described utilize a custom-designed device with a wearable computer and video cameras to tackle the issue of developing a navigational aid for the blind. From a research perspective, the advantage of having a device specifically built for this purpose is the ability to enhance the design more freely and change components as needed. This approach can more readily provide the solution that is better in terms of accuracy and efficacy. On the other hand, such an approach raises questions about the affordability of the device. Even with wearable computers and video cameras becoming increasingly affordable, without mass production of the custom device that integrates the two, it is likely to be relatively expensive and less affordable and therefore may not reach those in need. User preferences such as the level of verbosity in instructions provided by the device and the modality of instruction delivery are also important aspects to be considered. Although some work has been done on designing devices with variable levels of verbosity, no solution is yet available that is contextually aware of the needed level of verbosity. This would require monitoring a user's preferences and learning his or her needs based on a variety of factors, including whether the user is in a familiar environment.

With the increasing popularity and availability of smartphones, many technological services now have a greater reach. This applies to the deployment of indoor navigation aids for the B/VI population as it does to other technologies. Onboard sensors such as a gyroscope, an accelerometer, Bluetooth, and a Wi-Fi radio and NFC communication have made the smartphone a relatively low-cost platform for developing solutions for indoor navigation for B/VI users. This has ensured a future in which such solutions will be accessible to the masses in both the developed and the developing worlds. However, for the users to trust any indoor navigation aid and use it frequently, very accurate localization is required. For this reason, it is essential for any solution to be able to localize the user to within a meter of his or her actual location. Given the current trend in increased capabilities for smartphones, the smartphone-based solutions for providing navigation assistance to B/VI people seem especially promising.

Furthermore, any attempt at making indoor spaces more accessible to B/VI people has to be a concerted effort of innovators, lawmakers, and users. New accessibility standards may need to be introduced to improve the safety and independence of B/VI people navigating indoor environments. Promising technology solutions for navigation and wayfinding in these environments should inform these accessibility standards in the future.

## Acknowledgments

Preparation of this chapter was sponsored in part by the US Department of Transportation University Transportation Centers Program, Google Inc., and the National Science Foundation under NSF-NRI Award Number 1317989. Any opinions, findings, and conclusions or recommendations expressed in this material are those of the authors and do not necessarily reflect the views of any sponsors.

We were fortunate to benefit from the insights and expertise of several groups for this work. Colleagues from the Blind & Vision Rehabilitation Services of Pittsburgh (BVRSP) and the Western Pennsylvania School for Blind Children (WPSBC) have been instrumental in helping us to understand the needs and challenges of the B/VI population. Dr. George J. Zimmerman from the Vision Studies Program at the University of Pittsburgh provided invaluable insight into the field of orientation and mobility, which also significantly impacted this work.

The authors also thank the many members of the TechBridgeWorld research team at Carnegie Mellon University who provided assistance and insight that enhanced this work in many ways, notably, Dr. Balajee Kannan, Dr. M. Beatrice Dias, Ermine Teves, Sarah Belousov, M. Freddie Dias, Dr. Yonina Cooper, Nisarg Kothari, Dr. Gary Giger, Hend K. Gedawy, and Soyeon Hwang.

## References

Bhatlawande, S., Mahadevappa, M., & Mukhopadhyay, J. (2013). Way-finding electronic bracelet for visually impaired people. In Proceedings of *IEEE point-of-care healthcare technologies (PHT)* (pp. 260–263). New York, NY: IEEE.

Centre for Innovation and Development. (n.d.). Sensible blind canes: Electronically-enhanced mobility aids for the blind. Retrieved January 31, 2014, from http:// i4d.mit.edu/sensible-blind-canes/

Chintalapudi, K., Iyer, A. P., & Padmanabhan, V. N. (2010). Indoor localization without the pain. In *Proceedings of the sixteenth annual international conference on mobile computing and networking* (pp. 173–184). New York, NY: ACM. doi:10.1145/1859995.1860016

Espinosa, M., Ungar, S., Ochaíta, E., Blades, M., & Spencer, C. (1998). Comparing methods for introducing blind and visually impaired people to unfamiliar urban environments. *Journal of Environmental Psychology, 18*(3), 277–287.

Fallah, N., Apostolopoulos, I., Bekris, K., & Folmer, E. (2012). The user as a sensor: Navigating users with visual impairments in indoor spaces using tactile landmarks. In *Proceedings of the SIGCHI conference on human factors in computing systems* (pp. 425–432). New York, NY: ACM. doi:10.1145/2207676.2207735

Fallah, N., Apostolopoulos, I., Bekris, K., & Folmer, E. (2013). Indoor human navigation systems: A survey. In *Interacting with computers.* doi:10.1093/iwc/iws010

Galatas, G., McMurrough, C., Mariottini, G. L., & Makedon, F. (2011). eyeDog: An assistive-guide robot for the visually impaired. In *Proceedings of the 4th international conference on pervasive technologies related to assistive environments* (Article 58). New York, NY: ACM. doi:10.1145/2141622.2141691

Ganz, A., Schafer, J., Gandhi, S., Puleo, E., Wilson, C., & Robertson, M. (2012). PERCEPT indoor navigation system for the blind and visually impaired: Architecture and experimentation. *International Journal of Telemedicine and Applications, 2012,* Article ID 894869.

Giudice, N. A., & Legge, G. E. (2008). Blind navigation and the role of technology. In *The engineering handbook of smart technology for aging, disability, and independence* (pp. 479–500). Hoboken, NJ: John Wiley & Sons. doi:10.1002/9780470379424. ch25

How bone-conducting headphones work. (n.d.). Retrieved February 17, 2014, from http://electronics.howstuffworks.com/gadgets/audio-music/bone-conducting-headphones.htm

Hub, A., Diepstraten, J., & Ertl, T. (2003). Design and development of an indoor navigation and object identification system for the blind. In *Proceedings of the 6th international ACM SIGACCESS conference on computers and accessibility* (pp. 147–152). New York, NY: ACM. doi:10.1145/1028630.1028657

Kaiser, E. B., & Lawo, M. (2012). Wearable navigation system for the visually impaired and blind people. In *IEEE/ACIS 11th international conference on computer and information science (ICIS)* (pp. 230–233). New York, NY: IEEE. doi:10.1109/ ICIS.2012.118

Karlsson, N., Di Bernardo, E., Ostrowski, J., Goncalves, L., Pirjanian, P., & Munich, M. E. (2005). The vSLAM algorithm for robust localization and mapping. In *Proceedings of the 2005 IEEE international conference on robotics and automation, ICRA 2005* (pp. 24–29). New York, NY: IEEE. doi:10.1109/ROBOT.2005.1570091

Kay, L. (1964). An ultrasonic sensing probe as a mobility aid for the blind. *Ultrasonics, 2*(2), 53–59.

Kay, L. (2000). *Ultrasonic eyeglasses for the blind.* In proceedings of ASA/NOISE-CON 2000 meeting. Newport Beach, CA.

Kelley, P. (2009). Historical development of orientation and mobility as a profession. Retrieved February 4, 2014, from http://www.orientationandmobility.org/profession.html

Lahav, O., & Mioduser, D. (2004). Exploration of unknown spaces by people who are blind using a multi-sensory virtual environment. *Journal of Special Education Technology, 19*(3), 15–24.

Lahav, O., & Mioduser, D. (2008). Haptic feedback support for cognitive mapping of unknown spaces by people who are blind. *International Journal of Human-Computer Studies, 66*(1), 23–35.

Laoudias, C., Constantinou, G., Constantinides, M., Nicolaou, S., Zeinalipour-Yazti, D., & Panayiotou, C. G. (2012). The airplace indoor positioning platform for Android smartphones. In *Proceedings of the 2012 IEEE 13th international conference on mobile data management* (pp. 312–315). Washington, DC: IEEE Computer Society. doi:10.1109/MDM.2012.68

Lohmann, K., Yu, J., Kerzel, M., Wang, D., & Habel, C. (2014). Verbally assisting virtual-environment tactile maps: A cross-linguistic and cross-cultural study. In *Foundations and practical applications of cognitive systems and information processing* (pp. 821–831). Berlin, Heidelberg, Germany: Springer.

Lopes, S. I., Vieira, J. M. N., Lopes, Ó. F. F., Rosa, P. R. M., & Dias, N. A. S. (2012). MobiFree: A set of electronic mobility aids for the blind. In *Proceedings of the 4th international conference on software development for enhancing accessibility and fighting info-exclusion* (Vol. 14, pp. 10–19). doi:10.1016/j.procs.2012.10.002

National Federation of the Blind. (n.d.). GPS technology for the blind. Retrieved February 15, 2014, from https://nfb.org/Images/nfb/Publications/bm/bm06/bm0602/bm060206.htm

Picinali, L., Afonso, A., Denis, M., & Katz, B. F. G. (2014). Exploration of architectural spaces by blind people using auditory virtual reality for the construction of spatial knowledge. *International Journal of Human-Computer Studies, 72*(4), 393–407.

"Project Tango." (n.d.). Retrieved February 21, 2014, from http://www.google.com/atap/projecttango/

Quan, M., Navarro, E., & Peuker, B. (2010). Wi-Fi localization using RSSI fingerprinting. http://digitalcommons.calpoly.edu/cpesp/in/

Ran, L., Helal, S., & Moore, S. (2004). Drishti: An integrated indoor/outdoor blind navigation system and service. In *Proceedings of the second IEEE annual conference on pervasive computing and communications* (pp. 23–30). Washington, DC: IEEE Computer Society.

Ravi, N., Shankar, P., Frankel, A., & Elgammal, A. (2005). Indoor localization using camera phones. In *Mobile computing systems and applications, 2006. WMCSA '06. Proceedings. 7th IEEE workshop on* (pp. 1–7). New York, NY: IEEE. doi:10.1109/WMCSA.2006.4625206

Sánchez, J., Sáenz, M., Pascual-Leone, A., & Merabet, L. (2010). Navigation for the blind through audio-based virtual environments. In *CHI '10 extended abstracts on human factors in computing systems* (pp. 3409–3414). New York, NY: ACM.

Sheepy, E., & Salenikovich, S. (2013). Technological support for mobility and orientation training: Development of a smartphone navigation aid. In T. Bastiaens & G. Marks (Eds.), *Proceedings of world conference on e-learning in corporate, government, healthcare, and higher education 2013* (pp. 975–980). Chesapeake, VA: AACE.

Subryan, H. (2009). Tactile maps as navigational aids. Retrieved from http://udeworld.com/tactile-maps-as-navigation-aids

Torres-Solis, J., Falk, T. H., & Chau, T. (2010). A review of indoor localization technologies: Towards navigational assistance for topographical disorientation. In F. J. V. Molina (Ed.), *Ambient intelligence*. Rijeka, Croatia: InTech.

Trekker talking GPS. (n.d.). Retrieved January 30, 2014, from http://www.humanware.com/en-united_kingdom/products/blindness/talking_gps/trekker_extra_info/what's_new_in_trekker_2.7

Vítek, S., Klima, M., Husnik, L., & Spirk, D. (2011). New possibilities for blind people navigation. In *2011 international conference on applied electronics* (pp. 1–4). New York, NY: IEEE.

Wang, H., Sen, S., Elgohary, A., Farid, M., Youssef, M., & Choudhury, R. R. (2012). No need to war-drive: Unsupervised indoor localization. In *Proceedings of the 10th international conference on mobile systems, applications, and services* (pp. 197–210). New York, NY: ACM. doi:10.1145/2307636.2307655

Xu, X., & He, J. (2012). Design of intelligent guide vehicle for blind people. *Applied Mechanics and Materials, 268–270*, 1490–1493.

Yu, X., & Ganz, A. (2012). Audible vision for the blind and visually impaired. In *Engineering in Medicine and Biology Society (EMBC), 2012 annual international conference of the IEEE* (pp. 5110–5113). New York, NY: IEEE.

# 9

## The NavPal Suite of Tools for Enhancing Indoor Navigation for Blind Travelers

**M. Bernardine Dias**

### CONTENTS

*Abstract:* In this chapter, we describe our vision for computing technology tools that can enhance the safety and independence of blind and visually impaired (B/VI) people navigating unfamiliar indoor environments. We describe the overall road map for our work, along with the research to date on each component of this road map. This set of tools, known as "NavPal," combines a variety of techniques and technologies including robots, crowd-sourcing, advanced path planning, and multimodal interfaces. All of these tools and the framework have been informed, tested, and endorsed by many B/VI adults and by several orientation and mobility (O&M) experts.

### 9.1 Introduction

Assistive technology plays a key role in the independence and safety of people with disabilities. For B/VI people, appropriately designed and well-implemented assistive technology can additionally make a significant difference in education, social acceptance, and productivity. We have been working with B/VI communities around the world on a variety of assistive technology projects for almost a decade, with a more recent focus on technology tools for wayfinding and navigation. Given the availability of some tools to assist B/VI travelers with navigation in outdoor environments, our focus has primarily been on indoor environments and scenarios that necessitate transitions between outdoor and indoor environments. Because visual impairments can vary widely, the corresponding needs and preferences for tools and interfaces can also vary accordingly. Our work addresses this wide range of needs and preferences by exploring a suite of relevant tools that can accommodate several options for customization by users. Our approach also recognizes the current limitations of technology and the capabilities of the B/VI travelers by designing assistive mechanisms that value input from humans and therefore are essentially human-machine solutions. Another key element of our approach is a strong focus on needs assessment and iterative design with relevant user communities, as shown in Figure 9.1. Therefore, the problems we address in assistive navigation and the tools we design are heavily influenced by feedback from a variety of users in the B/VI community and relevant O&M experts.

Findings from our needs assessment revealed several concerns of B/VI travelers. First, they preplan their navigation routes as much as possible and need to build a mental map of the new environment they will be navigating. Once in a given indoor space, they need to orient themselves in that location so they can identify where they are at any given time. Next, they need to figure out how to navigate to and from the location(s) of interest from a known environment. They also need to be informed of dynamic changes to the unfamiliar environment that may impact their safe navigation. Furthermore, they need to be able to "record" their navigation experience for future trips and also potentially share this information with others who might find it

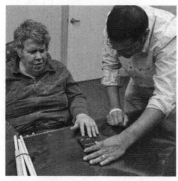  

**FIGURE 9.1**
Needs assessment and iterative user testing with B/VI people.

useful. Finally, if they get into any unsafe or difficult situations while navigating the unfamiliar environment, they need to have a reliable means of getting help. These findings support conclusions from previous work in this area and also align with the methodology adopted by O&M experts.

To orient themselves in an indoor environment and help identify their location, B/VI people rely heavily on landmarks and clues around them. Landmarks and clues can include elements such as doorways, changes in elevation, or distinct sounds and smells. During navigation, B/VI travelers depend on a mental or cognitive map of the space that they may have previously constructed and also use methods such as counting the number of landmarks or clues in their environment. If lost or disoriented, they typically ask for assistance from sighted people in the vicinity, although instructions from many sighted people are usually not specific enough to be useful. They also sometimes seek assistance from owners or managers of a specific facility they are visiting, O&M specialists (if it is a location they need to navigate frequently), or other B/VI people who are familiar with that location.

Effective indoor navigation and wayfinding tools should not only assist B/VI users to navigate indoors but also seamlessly integrate with outdoor navigation guidance, transit assistance, and other assistive services that enable B/VI travelers to safely and independently traverse urban environments. This means that urban navigation aids for this population must incorporate accessible interfaces that allow B/VI users to both receive and convey information, and they must be customizable to accommodate individual preferences. These tools must also be capable of indoor and outdoor localization at the resolution necessary for B/VI travelers. Access to maps and other information in a variety of forms will also be critical so that routes that adhere to sensory and other constraints can accordingly be planned. Finally, to truly empower B/VI travelers, assistive tools should provide mechanisms for advocacy to improve accessibility within the larger framework of the city infrastructure. This is because any attempt at making indoor spaces more accessible to B/VI people has to be a concerted effort of innovators, lawmakers, and users. For example, new accessibility standards may need to be introduced to improve the safety and independence of B/VI people navigating indoor environments. Promising technology solutions for wayfinding and navigation in these environments could help inform these accessibility standards in the future.

Our research in assistive technology over the past few years has been exploring how computing technology tools can enhance the safety and independence of visually impaired people navigating urban environments, with a focus on unfamiliar indoor spaces. Our work began with a smartphone application designed to provide dynamic localization and path-planning assistance to visually impaired people as they navigate indoor environments. This work soon evolved to address the more complete needs of the B/VI community in the area of indoor navigation. In this chapter, we describe the overall road map for our research in this area, along with the specific tools we have designed to

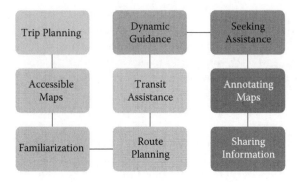

**FIGURE 9.2**
Necessary components of urban wayfinding and navigation for blind and visually impaired people.

date in each segment of this road map. This suite of tools, known as "NavPal," combines intelligent robotic systems, Wi-Fi-based signal strength estimation, constrained path planning, and multimodal interfaces toward improving the overall safety and reliability of navigation for a variety of B/VI users.

The envisioned road map to the NavPal suite of tools is illustrated in Figure 9.2. Each segment of the road map requires a set of assistive technology tools to enhance urban wayfinding and navigation for the B/VI community.

The goal of the NavPal suite of tools is to provide B/VI adults with customized guidance options during each phase of their travel in urban environments, beginning with trip planning, continuing through safe arrival at their destination, and ending with recording and sharing relevant information learned from each trip. Envisioned examples of NavPal components are a floor plan editing tool that allows building managers to easily create and edit floor plans in a format that can be easily made accessible to visually impaired people, enhanced route-planning algorithms that efficiently plan safe routes for B/VI users, and customizable user interfaces that effectively communicate route information with sufficient resolution to a variety of B/VI users. The NavPal suite of tools and the corresponding road map have been informed and iteratively designed with the input of many B/VI adults and several O&M experts. This work is strongly endorsed by all of these participants. The rest of this chapter describes the NavPal road map and its components in greater detail, with a focus on indoor wayfinding and navigation.

## 9.2 Trip Planning and Accessible Maps

Most B/VI people always plan trips in advance as much as possible. In fact, many blind people will not travel to unfamiliar locations unless it is possible to preplan most of the trip. Existing trip-planning tools do not take into account the variety of challenges and the complex components of trip planning that begin with leaving the home, do not allow customization for constraints and preferences of B/VI travelers, and are not able to accommodate a variety of the transportation modes commonly used by B/VI people. Therefore, an important tool envisioned for NavPal is an assistive trip planner that addresses these shortcomings. The NavPal trip planner will be one of the last components of NavPal to be designed and implemented, since it must integrate information from most of the other NavPal components. Therefore, we have not focused much effort on this tool to date. A highly relevant component of trip planning is having access to useful maps. For the B/VI population, this translates to accessible maps; that is, maps in formats that are accessible and have sufficient and relevant details for B/VI travelers.

A major obstacle for B/VI people preplanning trips is the lack of accessible maps for indoor environments. A survey of property managers revealed that the primary reason for this roadblock is the lack of tools to create and maintain accessible maps of the facilities they manage. Furthermore, there is no single authority or commercial solution that maps indoor areas like the widely available outdoor map equivalents. Therefore, a solution to creating accessible indoor maps is likely going to take a different form than one for its outdoor counterparts.

Toward further understanding the challenge of accessible map creation, we interviewed five building managers from a mix of for-profit and nonprofit organizations about the variety of difficulties that can arise in making their locations accessible to B/VI visitors. Most buildings have one or two main entrances, so they employ a front desk attendant to provide directions to visitors and to prevent unauthorized visitors from entering the location. These front desk attendants are often instructed to warn B/VI visitors about obstacles that they might have difficulty avoiding unless they are escorted by someone who can guide them through the building. While many buildings have room numbers displayed in braille, they do not have braille directional signs to help guide B/VI visitors. An additional complication is that many B/VI people do not read braille. Special events that alter the regular configuration of an environment, construction or maintenance, and architectural features such as spiral stairways can cause additional navigational difficulties for B/VI visitors. In addition, floor plans are not always maintained accurately as buildings undergo alterations.

In general, B/VI visitors are expected to be able to navigate buildings by themselves or bring their own guidance assistants with them when they visit most indoor environments. Most building managers and front desk attendants also have limited knowledge about the challenges of B/VI people and are not always able to assist in useful ways. All of the building managers we interviewed agreed that a tool for creating and maintaining accessible maps would be very helpful. Important features in an accessible map creation tool identified from these interviews are briefly described as follows:

- *Feature classification:* the ability to classify rooms, halls, doors, and types of doors
- *Annotation of connectors:* the ability to annotate doors, elevators, regular exits, and emergency exits
- *Naming areas:* the ability to label spaces at different granularities (e.g., office numbers, building names)
- *Map updates:* the ability to easily update maps when the environment changes or an error is discovered
- *Using existing blueprints:* the ability to import images or blueprints of floor plans in a variety of file formats and edit them as needed
- *Automated information extraction:* automated information extraction from imported blueprints where possible and the ability to easily edit the resulting maps
- *Privacy control:* the ability to designate areas of maps as private and make other areas of maps publicly accessible or accessible to specific people
- *Effective interface:* an intuitive interface that is easy to use
- *Automated accessibility feature insertion:* automated conversion of resulting maps to information and formats accessible to B/VI people

We developed an initial prototype for a floor plan creation and management tool that can better equip building managers to create and maintain accessible maps of their buildings and surroundings. This initial prototype was implemented under our guidance by a group of students in the "Software Development for Social Good" course at Carnegie Mellon University. The primary menu for this tool is shown in Figure 9.3. The tool allows users to upload existing images of floor plans, which are converted to an editable digital format. Image processing techniques are used to identify and label

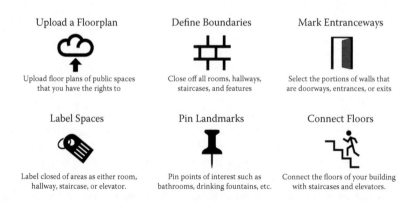

**FIGURE 9.3**
Initial prototype of floor plan creation and management tool.

relevant features. First, line extraction identifies all straight edges in each floor plan image. These lines that are in close proximity to one another are then merged to reduce errors. The resulting automated identification of walls is then further modifiable by the user. After line identification, the autodetect process identifies square areas that could indicate where doors are potentially located. Feature detection is then used to isolate and identify potential doors in the floor plan image. Finally, text recognition is used to extract any labeled areas (such as room numbers and building names). The text is first isolated by identifying conjoined lines, and then Tesseract is used to perform optical character recognition (OCR) on the isolated text.

This automated process of extracting information from the uploaded floorplan images dramatically reduces the amount of work the user has to do to convert existing floor plan images into accessible maps. In addition, the prototype tool allows users to draw new walls, complete rooms, label areas, add doors, and perform other typical edit functions. Initial user feedback from building managers on this prototype was very positive, and we are currently creating a more enhanced prototype that incorporates initial user feedback. Our initial methodology for creating accessible maps is to convert them into a format that can be interpreted by the other NavPal tools and thus be used by B/VI people via the accessible interfaces on these tools. However, we are also working with relevant experts to explore ways in which these floor plans can be made more accessible via screen readers and other currently ubiquitous tools. Next we describe our vision for how accessible maps can be used by B/VI travelers to familiarize themselves with indoor locations and plan effective routes through these spaces.

## 9.3 Familiarization and Route Planning

Once we have accessible building maps, we need tools that allow B/VI people to familiarize themselves with these environments so that they need fewer interventions from others when they travel to unfamiliar locations. Depending on how frequently a B/VI person needs to visit an unfamiliar location, he or she may get help from a friend, an information line or desk at the location, a sighted person in the vicinity, or an O&M specialist as needed to construct a corresponding mental map and learn how to navigate that environment. While we do not envision humans being completely removed from this process, technology can certainly play a greater role in enhancing the independence of B/VI people in this familiarization. Toward this end, we have begun to explore several options for developing a familiarization tool that can take advantage of the accessible maps generated by the NavPal tool described in the previous section and that interconnects with the other components of NavPal, including the route-planning tool and trip-planning tool. The most promising avenue for this familiarization tool is the use of virtual environments.

Our vision for the NavPal familiarization tool is to transform the accessible digital maps into accessible virtual environments that B/VI users can explore to build detailed cognitive maps of each environment before physically interacting with that environment. The virtual environment can be represented using audio cues and sounds associated with objects in the environment. In addition, force-feedback via a joystick and vibrations can be used to guide the B/VI users and alert them to obstacles and objects of interest in the virtual environment. We are currently exploring methodologies for constructing these virtual environments from accessible digital maps and effectively enabling B/VI users to familiarize themselves with these environments.

Once visually impaired travelers have familiarized themselves with an environment, they often like to plan routes within that environment that will allow them to effectively traverse between key locations such as the entrances and exits to the building, restrooms, elevators, cafeterias, meeting rooms, and so on. Several aspects are notably different for planning routes for visually impaired adults than for route planning for sighted adults. For example, the routes need to be optimized for different objective functions such as minimizing turns, maximizing traveling through areas where others may be present in case the traveler needs assistance, favoring paths where the traveler can follow a wall closely, and ensuring the chosen routes have as many accessible landmarks and clues as possible. To learn more about these route-planning needs and preferences of B/VI people, we created a simple Web tool for B/VI travelers using screen readers to preplan routes within a small set of indoor environments.

This NavPal routes Web tool (shown in Figure 9.4) provides the user with the option of selecting a start and destination from the available sample locations and provides route information for walking between the two locations. The user is first allowed to select a building and then select a location within that building as the start and goal locations. Buildings and locations within each building are provided via drop-down lists. The Web tool is designed to be accessible for B/VI users via screen readers and was tested with some of the most commonly available screen readers. We further tested the accessibility of the tool using WAVE, the Web accessibility evaluation tool provided by WebAIM. For the purpose of this initial investigation of needs and preferences, we used fictitious route information. However, several route-planning algorithms are available to plan effective routes given specified user constraints and preferences. Before exploring the route-generation aspect of this work, we wanted to understand the interface constraints and how useful this tool will be to the target population. A survey of B/VI adults confirmed that this tool will be very useful and provided several insights into the range of needs and preferences that must be taken into consideration when converting this prototype into the final NavPal route-planning tool. We are currently implementing

**FIGURE 9.4**
Initial prototype of an accessible route-planning tool for indoor environments.

this tool and expect it to be able to use the NavPal-accessible maps and provide input to the NavPal-dynamic guidance tool described in the next section.

## 9.4 Transit Assistance, Dynamic Guidance, and Seeking Assistance

The role of computing tools is not limited to trip and route planning. Assisting visually impaired travelers with dynamic guidance during transit and indoor navigation and with seeking assistance during the trip is the next critical challenge. These tools can help in situations where unforeseen dynamic situations arise (e.g., an elevator being out of operation) and also allow for reminding the traveler about aspects of the route that he or she has forgotten. In addition, if needed, the tools can assist the traveler in seeking assistance from others in the environment or guiding the traveler to safety. It is essential to view these tools as human-machine systems since it is important to ensure that the traveler is able to use his or her orientation and navigation skills in this process.

While transit assistance is usually considered outdoor navigation or assigned to a special category, there are indoor components to transit that are important, and hence we include transit assistance in this description of the NavPal suite of tools. This tool is useful, for example, when navigating large transit stations such as airports and train stations. We are currently investigating the needs and constraints that must be considered when

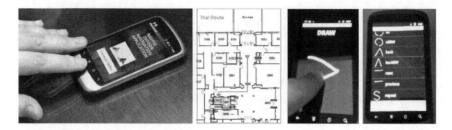

**FIGURE 9.5**
Initial prototype of the NavPal dynamic guidance tool implemented on a smartphone.

designing a tool that can assist B/VI travelers with navigating the unique aspects of transit stations.

To address dynamic guidance during indoor navigation in more general locations, we developed an Android smartphone tool, shown in Figure 9.5, that integrates indoor localization (Kothari, Kannan, Glasgow, & Dias, 2012), sparse map representation (Kannan et al., 2014), and an accessible user interface (Gedawy, 2011). Specifically, we developed a navigation solution that combines dead reckoning and Wi-Fi signal strength fingerprinting with enhanced route-planning algorithms to account for the constraints of B/VI users to efficiently plan routes and communicate the route information with sufficient resolution. The localization component of this initial work used a small wheeled robot to initially map the indoor environments and build a database of Wi-Fi fingerprints. This P3DX robot was retrofitted with a laser range finder for obstacle detection and mapping and a fiber-optic gyroscope for localization. The robot was remotely operated to roam a building and carried a smartphone and thereby constructed a Wi-Fi signal strength map that corresponds to the building map generated by the robot's sensors. The smartphone app was then able to use this Wi-Fi map to localize the user during navigation. The interface component of this prototype used simple on-screen gestures and a combination of voice and vibration feedback to allow B/VI users to interact with the tool. To evaluate the feasibility of our solution, we developed a prototype application on a commercial smartphone and tested it in a small sample of indoor environments.

The map of an indoor environment was represented in this prototype using a variation of hierarchical maps to accommodate dynamic changes while maintaining a compressed representation suitable for a smartphone (Kannan et al., 2014). In this representation, indoor locations were represented on a map as nodes on a graph, and the map was split into subgraphs. A variation of the D* algorithm was used to efficiently plan and replan routes dynamically despite the limited computing power available on the smartphone (Kannan et al., 2014).

More specifically, in this hierarchical map representation, low-level maps used for higher resolution navigation within rooms and hallways represented individual rooms with significant spatial detail without having to represent the spatial relationships to other rooms. Complementarily, high-level maps used to generate plans for coarse navigation between floors and rooms represented larger areas of a building while omitting detailed spatial relationships of individual locations inside rooms and corridors. In this pro-totype, high-level maps were represented as graphs and low-level maps as grids. The high-level route planner first searches for an optimal path on the graph and provides a restricted set of nodes to the low-level route planner. This grid planner then traverses the provided nodes and generates a higher resolution path to the destination.

The interface for this prototype supported three levels of verbosity for rout-ing instructions, as illustrated in Figure 9.6. In the high-level option, the user is guided at the hallway intersection level until he or she gets to the final hallway, at which point a count of doors along the way to the destination is given. The intermediate level adds step-by-step directions along with a count of doors along the whole route. The low level adds additional landmarks and contextual information about the environment. The interface also used vibration patterns to provide tactile feedback. The navigation menu allowed the user to perform tasks such as specifying the destination, getting route directions, and exiting the application. On-screen gestures were used to instruct the tool to repeat an instruction, go back to the previous instruction, or repeat all instructions.

Since we do not envision technology solutions completely eliminating the need for seeking human assistance in the future, it is important to design tools in a manner that enables B/VI users to seek assistance from available sources when needed. For example, a graphical map view on the dynamic guidance tool enables sighted bystanders to provide assistance to the B/VI traveler more easily since the bystander can simply click the relevant location on the map to indicate a place of interest. We are also exploring other ways in which the NavPal tools can more effectively enable B/VI users to seek assistance when necessary.

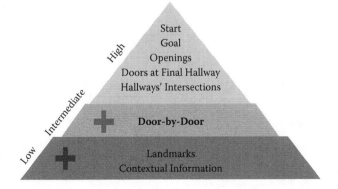

**FIGURE 9.6**
Examples of wayfinding information that could be presented to B/VI users at three levels of verbosity.

## 9.5 Annotating Maps and Sharing Information

For longer term usefulness and sustainability, it is important to enable B/VI travelers to influence maps both on a personal basis (annotations) and in a public manner (to share information and advocate for changes). To explore these topics, we did some initial enhancements to our NavPal dynamic guidance tool and experimented with an initial crowdsourcing app, as shown in Figure 9.7. First, we added landmark and obstacle input options to the NavPal dynamic guidance tool. In the initial prototype, the "landmark" option allowed the user to annotate the "current" location on the map with a text or audio label for a landmark. The "obstacle" option allowed the user to add obstacles to the map that would be recognized by the route planner in future iterations. This option allows the user to specify the object's location, its direction relative to the user's locations, and its width and height in feet by choosing one of three options:

- *Here:* Choosing this option directs the interface to retrieve the current location. The user is then forwarded to the directions menu to specify the object's direction relative to the user's position. Then the user gets to specify the width and height of the object.

- *Remote relative:* This option can be useful if the user is made aware of or detects an object from a distance and can estimate the relative distance to it. As in the previous option, the user is then forwarded to specify the direction, width, and height of the object.

- *Remote absolute:* This option allows a user to directly insert an object at a specific absolute location in the global coordinate frame. As in the previous two options, the user then gets to specify the direction, width, and height of the object.

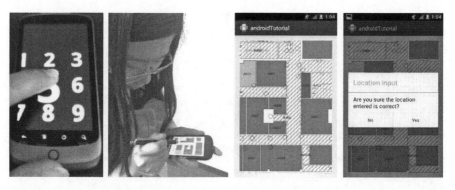

**FIGURE 9.7**
Initial explorations with annotation and crowdsourcing.

For numerical inputs such as width and location coordinates, a modified version of the "Talking Dialer" was used. The Talking Dialer is a free Android application that has a virtual number pad. Wherever the user puts his or her finger on the screen, the number 5 appears, and the rest of the numbers are spread relative to number 5. The user feels a vibration and hears a buzz as he or she transitions between the numbers. When the intended number is reached, the user lifts his or her finger off the screen. This action selects the number, speaks it out loud, and prints it on the screen. Shaking the phone deletes the last number in the list of selected numbers. To confirm input, the user clicks the trackball twice. The first click speaks out loud what the user entered to double-check with the user that this is correct. The second click confirms the entry. The Talking Dialer was modified for NavPal in several ways. First, the dialing ability was disabled. Second, the instructions were modified to reflect the new-use context. Third, the application confirmed each value entered by the user.

Toward the goal of sharing relevant information, we also explored an initial crowdsourcing approach for coarse indoor mapping of Wi-Fi signal strengths in buildings. Since a significant percentage of the adult population in the United States now owns a smartphone, crowdsourcing indoor localization information can be a viable option. Moreover, since crowdsourcing does not require any costly infrastructure installation, it is also an affordable solution. We implemented a simple mobile application that displays a map interface showing the floor plan of the current location of the user. The study participants were asked to mark their location on the map as they traversed the building. Each time a point was marked on the map, the mobile application recorded the coordinates of the point on the map along with details of all the Wi-Fi access points (APs) detected by the phone at that location. Each of these points along with their associated Wi-Fi data can be used as indoor landmarks. Through the crowdsourcing, a database of indoor Wi-Fi landmarks was created, which was then used to localize test points.

Each entry in the database is represented in the form of a vector. Each element in the vector represents the RSSI of a detected AP, and the detected APs are considered in the same order for each point so that they can be compared. A nonzero element indicates that the corresponding AP was detected by the Wi-Fi scan at that point on the map and the value of the element is the RSSI of that AP. A zero element indicates that a particular AP was not detected. Each of these vectors has corresponding "real" coordinates (ground truth) on the map that was input by participants via crowdsourcing. To localize a point during a test run, the vector of the point is compared with the vector corresponding to each point in the database to find the most similar database entry. This comparison is done by calculating Euclidean distances between the vector representing the test point and each vector representing points in the database set.

## 9.6  Conclusions and Design Guidelines

Our research in assistive technology over the past few years has explored avenues to enhance the safety and independence of B/VI people navigating indoor urban environments. Our work began with a smartphone application designed to provide dynamic localization and path-planning assistance to this community and evolved to address the more complete needs of the B/VI community in the area of indoor navigation and wayfinding. In this chapter we described a road map for our research in this area, along with the specific tools (collectively named NavPal) we have designed to date in each segment of this road map. We conclude this chapter with a set of 10 design guidelines we have discerned from this work to date that will be informative to other researchers working in the area of assistive navigation technology for B/VI users.

1. *Include users in the design process:* The most important guideline for effective work in this area is to include B/VI users and O&M specialists in an iterative design and testing process. Testing conducted with blindfolded sighted users does not yield the same result.

2. *Keep the user in the loop:* It is important to appreciate the O&M training of the B/VI users. Navigation assistance tools that keep B/VI users in the loop and incorporate their input effectively will produce more robust and useful guidance.

3. *Pay attention to affordability:* Employment and purchasing power can be low among the B/VI community, so assistive tools must be affordable in order to be useful.

4. *Reliability is extremely important:* Because the failure of these tools could jeopardize the safety of B/VI users, the tools must operate reliably to gain the trust of the users. Reliability can be addressed practically by predictable behavior in all scenarios and graceful degradation in difficult environments. If users can predict failure conditions and how the tool will respond to these conditions, users can detect these situations and be prepared to overcome these limitations of the technology.

5. *Build practical tools:* It is important to ensure that B/VI users will be able to learn to use the technology in a reasonable time frame and that the usage scenario of the tool is practical. For example, practical considerations such as theft of expensive technology or the inability of a user to carry and use many, heavy, or poorly shaped devices must be taken into account when designing these tools.

6. *Do not overwhelm the user:* Because the B/VI users must pay attention to their surroundings and keep track of a variety of things when navigating, it is important to design assistive tools that do not overwhelm

or monopolize the attention of the user. Interfaces should be as simple as possible and allow for customization since the user will often need to multitask and will want different levels of assistance from the assistive device depending on the specific scenario.

7. *Environmental considerations are important:* In areas with high levels of competing sound, B/VI users often prefer less information from assistive tools so that they can focus on other inputs from the surroundings. In contrast, when navigating through a large empty space, the user may want much more detailed instructions. Furthermore, when a tool provides information that is read aloud, it is important to consider privacy issues and whether this output is generating distracting levels of background noise.

8. *Expect and adapt to dynamics:* Effective assistive tools will provide mechanisms for recognizing changes in the environment and adapt to those changes in a timely manner that is beneficial to the B/VI user. Distinctions should be made between temporary changes and permanent infrastructural changes for optimal performance.

9. *Make the most of existing resources:* Understanding how B/VI users navigate without technology and employing universal design principles to harness resources useful to sighted people can contribute significantly to the success of assistive tools for the B/VI community.

10. *Understand the bigger picture:* Understanding procedures, policies, and laws relevant to accessibility can significantly contribute to successful design decisions for assistive tools.

## Acknowledgments

Many individuals and organizations supported the work presented in this chapter. Preparation of this chapter was supported by the US Department of Transportation University Transportation Centers Program, Google Inc., Boeing, Carnegie Mellon University's Berkman Faculty Development Fund, the National Science Foundation under NSF-NRI Award Number 1317989, and the National Institute on Disability and Rehabilitation Research under the RERC-APT (Grant Number H133E080019). Any opinions, findings, and conclusions or recommendations expressed in this material are those of the author and do not necessarily reflect the views of the sponsors.

Partner involvement has been essential for this work, and we were fortunate to collaborate with several groups who have supported various aspects of this project. Colleagues from the Blind & Vision Rehabilitation Services of Pittsburgh (BVRSP) and the Western Pennsylvania School for Blind Children (WPSBC) have provided guidance on this work and have encouraged their networks to participate in our needs assessment and testing. Dr. George J. Zimmerman from the Vision Studies Program at the University of Pittsburgh provided invaluable insight into the field of orientation and mobility, which also significantly impacted this work.

The author also thanks the many members of the research team who assisted with different aspects of this work, notably, Dr. Balajee Kannan, Hend Gedawy, Nisarg Kothari, Evan Glasgow, M. Freddie Dias, Sarah Belousov, Ermine Teves, Dr. David Kosbie, Daniel Muller, Vansi Vallabhaneni, Zhiyu Wang, Paul Davis, Justin Greet, Maxime Bury, Dr. M. Beatrice Dias, Dr. Yonina Cooper, Syed Ali Hashim Moosavi, Anna Kasunic, Alekhya Jonnalagedda, Ming Wu, Lucy Pei, Satish Ravishankar, Dr. Gary Giger, Sam Jian Yu Li, Hannah Flaherty, and Dr. Aaron Steinfeld.

## References

Gedawy, H. K. (2011). *Designing an interface and path translator for a smart phone-based indoor navigation system for visually impaired users* (Unpublished master's thesis). Carnegie Mellon University, Pittsburgh, Pennsylvania.

Kannan, B., Kothari, N., Gnegy, C., Gedawy, H., Dias, M. F., & Dias, M. B. (2014). Localization, route planning, and smartphone interface for indoor navigation. In A. Koubaa & A. Khelil (Eds.), *Cooperative robots and sensor networks* (pp. 39–59). Berlin, Heidelberg, Germany: Springer.

Kothari, N., Kannan, B., Glasgow, E. D., & Dias, M. B. (2012). Robust indoor localization on a commercial smart phone. *Procedia Computer Science, 10*, 1114–1120.

# 10

## Future Directions in Indoor Navigation Technology for Blind Travelers

M. Bernardine Dias

Aaron Steinfeld

M. Beatrice Dias

## CONTENTS

*Abstract:* This chapter examines several interesting directions for future research related to indoor wayfinding and navigation tools to assist blind travelers. In the short term, limitations of localization techniques can be improved by incorporating a variety of sensors and crowdsourcing into the localization framework. In addition, user interfaces can be improved with simpler gestures, clearer audio feedback, enhanced landmark lists, and more customizable instructions to enhance accessibility. Finally, effective representations of large-scale floor plans, appropriate maps, and efficient path-planning algorithms must be explored for enhanced route planning.

The grander vision for accessible navigation solutions will certainly involve a more systematic change in the general structure of urban environments.

Building smarter cities that are truly accessible to all is the ultimate goal. Several factors will contribute to realizing this vision. As robotics technology evolves to a stage where robots become ubiquitous, we need to ensure that they are equally capable of interacting with humans with disabilities. To this end, meaningful human-robot interaction methodologies must be developed to enable assistive robots for blind travelers. Smartphones and other mobile devices will be the primary modalities for information access and exchange, while mobile and stationary robot assistants could be utilized to help when needed. By identifying these short-term and long-term research goals, we hope to encourage researchers around the world to contribute their talents and efforts to enhance the growing literature on assistive technology and create a wider variety of useful solutions that will enable visually impaired people to safely and independently navigate indoor environments.

## 10.1 Introduction

According to the World Health Organization (2012), over a billion people worldwide have some form of disability. Often these disabilities prevent people from becoming active and independent members of society. As the elderly population grows globally, the number of people with disabilities rises accordingly, and issues of accessibility have increasingly important social and economic consequences. These consequences also impact disabled war veterans and other trauma survivors. Technology must play a critical role in alleviating these consequences and ensuring that people with disabilities enjoy greater independence. In the United States, the recent White House Technology Showcase (Dale & Deutchman, 2010) celebrating 20 years of the Americans with Disabilities Act highlighted the need for using technology to enable Americans with disabilities to participate fully, in both their personal and their professional lives. A critical component of this envisioned independence for people with disabilities is their ability to travel to unfamiliar locations and effectively navigate urban environments.

For visually impaired people, travel can be challenging and often daunting to the point that they avoid unfamiliar environments when possible. Since most environments are constructed to be easily navigated by sighted people who can walk, people with disabilities have to often seek help and use secondary cues to travel effectively. For example, although information such as room numbers might be encoded in braille in modern buildings, the vast majority of the information provided to sighted people remains inaccessible to many who are visually impaired. Day-to-day activities such as shopping and using transit systems remain challenging tasks for people with visual impairments. Some aids used by visually impaired travelers when navigating indoor environments are shown in Figure 10.1.

**FIGURE 10.1**
Example indoor navigation aids for visually impaired travelers (left to right: white cane, braille encoding for a map and room number, braille signage for a bathroom and elevator).

Worse still, emergency situations necessitating evacuation from an unfamiliar building is one of the greatest fears harbored by visually impaired people. Today, cities around the world are being planned and enhanced with the goal of ensuring accessibility to everyone, including people with disabilities. However, if these smart cities of the future are to be truly accessible to all, they must address the variety of navigation challenges faced by people with disabilities.

As robotic technology becomes more available and embedded in our day-to-day activities, one can anticipate that these tools will further enhance the navigation experience for visually impaired persons. Potential scenarios include the following:

- Robots assisting humans to localize within necessary resolution and context using a combination of perception, robot localization, and crowdsourcing
- Robots assisting humans to retrieve lost or fallen objects or to locate objects or people of interest
- Robots assisting humans to interact with other aspects of the environment such as reading notices
- Robots assisting humans during emergency evacuation of buildings

To this end, addressing the assistive human-robot interaction challenge must include areas of communication that encompass the ability to exchange information as well as learn and teach, interaction mechanisms that include mobility and manipulation, and coordination mechanisms that allow for a wide range of connectedness among cooperating heterogeneous agents. This chapter explores these possibilities for the future of technology solutions as they apply to visually impaired travelers navigating dynamic and unfamiliar indoor environments.

## 10.2 Short-Term Possibilities

Advances in assistive technology have made progress in enabling visually impaired travelers to more easily navigate public transit stations and various other indoor environments, as well as adapt to dynamic changes in those settings. Trip-planning tools (e.g., maps, transit options), crowdsourcing assistance, and accessible user interfaces have all contributed to this growing area of innovation. However, there are still many limitations associated with these technologies, and several indoor navigation challenges faced by visually impaired travelers remain unresolved. Potential enhancements that can be made in this area of research in the short term are explored in this section.

### 10.2.1 User Interfaces

Public transit is critical for employment, quality of life, and independence for people with disabilities (Steinfeld, Zimmerman, Tomasic, Yoo, & Aziz, 2011). Transportation settings also provide an interesting domain to test indoor navigation technology because of the distributed physical environment, dynamic complexity, and requirement for good navigation. This is especially true for larger transit stations where multiple floors, large crowds, and convoluted paths are often the norm. Wayfinding in transit stations can be challenging even for sighted users. Traditional guidance systems do not provide continuous and dynamic orientation information, which is desired by people who are blind when traversing large spaces like train station lobbies and complex intersections (Morton & Yousuf, 2011). Also, when supporting information is provided (e.g., maps or timetables), it is often not in an accessible form. For example, information kiosks in transit stations are required to be accessible by regulations, but compliance is uneven, particularly with electronic interfaces.

Maps play a critical role in guiding travelers. However, these maps may not be simple floor plans but instead be more complex representations that can include floor maps, road maps, transportation options, rail maps, accessibility information, and more. Representing this graphically rich information in accessible formats is an open research problem. There are well-established guidelines and approaches for making such systems accessible (e.g., Trace Center, 2000), and alternative formats can be obtained. Traditional tactile maps and three-dimensional models, including approaches with digital information layers (Kehret, Miele, & Landau, 2011), can be used effectively but are limited to features in the environment that are permanent. Remote human help on demand is sometimes available and can be found in various forms. This can be the classic white courtesy phone or help buttons with speakers. Unfortunately, all of these approaches are physically anchored in a static location and lack the ability to gesture in the direction of needed travel. Future research in this area will benefit from a variety of topics in the literature, including automating tactile graphic production, using sound to represent distance, using vibrations and sound to help blind people experience graphics, narrating map information, using text labels for graphics that can be read via screen readers, and more (e.g., Kim & Zatorre, 2010; Su, Rosenzweig, Goel, de Lara, & Truong, 2010; Wall & Brewster, 2006).

The quality of user interaction with a system often dictates its usefulness. Buttons, gestures, and speech are common modalities for input by blind users (Mau, Melchior, Makatchev, & Steinfeld, 2008). Most electronic mobility aids for visually impaired users rely on audio-based interfaces, using Text-To-Speech (Tissot, 2003) or sounds as output mechanisms to direct and inform the user about routes and locations. Some systems also use vibration patterns, iconic sounds, metaphors, spatial sounds, spatial language, and virtual displays for the interface component (Klatzky, Marston, Giudice, Golledge, & Loomis, 2006). However, the interaction modality is often predetermined, and the user does not have the flexibility to easily switch between different modalities depending on the situation. Accessible interfaces (Way & Barner, 1997) is a topic of interest for many applications, and hence recent work has explored a variety of relevant topics, including using audio (Kane, Bigham, & Wobbrock, 2008), on-screen gestures (Kane et al., 2008), vibration patterns (Jayant, Acuario, Johnson, Hollier, & Ladner, 2010), multitouch options (Kane et al., 2008), methods for efficiently navigating large touch screens (Kane et al., 2011), and even methods for using sound to display distance information to blind travelers (Talbot & Cowan, 2009).

Previous work in this area (Kannan et al., 2014) also revealed that a key characteristic of accessible interfaces is maintaining user flexibility in terms of the verbosity of presented information and the different input and output modalities. The interface should be easily customizable to provide a good balance between the quality of the navigation instructions and the automatic generation of these instructions. For example, different levels of routing instruction details could give the user an option of choosing to be guided with information about the hallways and intersections, step-by-step directions, a count of doors, additional landmarks and contextual information about the environment, and so on. The interface should also support different input and output modalities, including visual, audio, gestures, and haptic feedback. Another important factor is to enable the user to add and delete annotations. The annotations could be obstacles, landmarks, reminders, and so on. It is also important to provide multiple output mechanisms. Audio output augmented with haptic or vibrational feedback has proved useful in past work (Kannan et al., 2014), especially in crowded environments.

Many people who are blind or low vision have strategies for hearing voice interactions from their phones. This typically manifests as speaker-phone interaction (e.g., like Siri) or with an ear bud in a single ear. The latter will often be removed and reinserted as needed, since covering an ear can impair spatial awareness. There have been attempts to explore alternative audio inputs, but this requires niche hardware (Marston, Loomis, Klatzky,

Golledge, & Smith, 2006). Likewise, there has been considerable research on successful audio methods for providing navigational guidance (e.g., Gaunet & Briffault, 2005; Rice, Jacobson, Golledge, & Jones, 2005). One of the more challenging aspects of designing user interactions for people with disabilities is to limit dependence on niche hardware. While such equipment is often very effective (e.g., Miller, 2012), it typically requires the user to carry an additional piece of equipment that is difficult to quickly replace or service. Also, dedicated hardware can often be expensive or become orphan technology because of small market factors. Smartphones are appealing alternatives because of their ability to serve as a relatively low-cost multifunction device and the ease with which they can be replaced and reconfigured. For example, apps can be redownloaded to a new phone, and many store their preference data in the cloud.

Future work in this area must explore a variety of techniques and modalities to develop recommended specifications for accessible interfaces in the context of indoor navigation. Some key topics in this research area are depicted in Figure 10.2.

Critical aspects of accessible user interfaces that must be explored and understood are the best way to present specific types of navigation information; the best way to obtain input from visually impaired users, trade-offs between customizable features and core components, and interface components that allow visually impaired users to interact with sighted users in the context of urban navigation (e.g., when seeking help); and, perhaps most

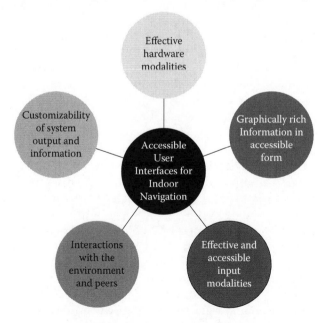

**FIGURE 10.2**
Key topics that need to be addressed in future work for accessible user interfaces.

critically, the best way to represent all of this information in a suitable world model that enables effective navigation. More work is also needed to investigate how an accessible smartphone interface might be used to remotely navigate other interfaces (such as touch-screen interfaces on stationary information kiosks) to access relevant information in a variety of contexts.

New, mainstream wearable technologies are also providing novel opportunities to introduce different interfaces. Developers have already implemented crowdsourced picture descriptions in Google Glass,* and many smartwatches have vibration capability. The bone induction speaker in Glass is also a convenient way to retain audible interaction without the user constantly inserting and removing an ear bud. Unfortunately, Glass is expensive, and there has been at least one case of a bystander grabbing a unit off a user's face. Some cochlear implant users have reported similar events when thieves mistook their behind-the-ear processers for a Bluetooth headset. While Glass is an attractive theft target and potentially unwelcome in certain venues, smartwatches may provide a middle ground. Watch vibrations are particularly interesting as a method for alerting users to information about key route decision points and navigation errors (e.g., Marston, Loomis, Klatzky, & Golledge, 2007). Watches linked to smartphones also allow users to quickly and subtly enter commands such as recalculate route, pause instructions, and repeat last spoken direction.

### 10.2.2 Crowdsourcing

An expert report initiated by the Office of Science and Technology Policy (Geo-Access Challenge Team, 2011) identified a strong need for integrated information: riders with disabilities want the ability to link points of interest, travel data, and municipal infrastructure in order to develop a full picture of what they might expect when executing a trip. Unfortunately, there are very little accessibility data stored in municipal databases. Therefore, most of the deployed systems that provide accessibility information in this context are based on crowdsourcing (e.g., IBM Accessibility City Tag, Wheelmap, CitiRoller, and OpenStreetMap). When designing crowdsource systems that serve people with disabilities, designers must provide value to all users, rather than simply rely on altruistic and ownership-induced behavior (Steinfeld et al., 2011). Furthermore, the appearance and language of the tools must often be designed to embody universal design principles so it would appeal to a wide audience. Wide appeal is critical in crowdsource systems, since performance usually improves as the user base grows. However, there are cases where a universal design approach to crowdsourcing will not work because of the nature of the task.

In many cases, a lost or disoriented building visitor encounters problems upon incorrectly traversing a hallway intersection or entry door. The most

---

* See http://www.engadget.com/2013/08/02/dapper-vision-openglass/.

easily recoverable scenario is when the user realizes there is a problem and pauses before committing a turn error. Many people already employ an effective, low-tech form of crowdsourcing by asking a bystander for clarification. This approach, often called the "nearest warm body" method, leverages the bystander's local knowledge and perception abilities to discriminate between confusing options or replace forgotten information. There are two problems with relying on bystanders. First, this method requires the presence of willing bystanders, which may be an issue in less traveled parts of buildings or in foreign language settings. Second, there is often no method to identify or mitigate incorrect information. The latter is similar to the case where visitors do not realize they made a navigation error before realizing they need help. Indoor smartphone navigation systems can help mitigate navigation errors and detect problems. However, these systems do not leverage human perception or crowd data to identify and address trouble spots. For example, a route may specify the use of elevators but not clarify that the leftmost elevator does not serve the desired floor. Information about subtle trouble spots can be accumulated from the crowd and inferred from navigation traces by earlier users.

Successful approaches will require management of three system design factors, which are shown in Figure 10.3. First, crowd-powered approaches need user input. Universal design is a proven way to expand value to all users, not just those willing to help people with blindness or low vision. Disorientation and navigation help has universal value, so it should be possible to create a crowd experience that attracts users without disabilities. Second, certain information has a short life span. The elevator example illustrates the case where the contributed data have persistent value and can be used indefinitely. Examples where the data rapidly become stale include maintenance detours, emergency evacuation, and congestion due to irregular events (e.g., parades, festivals). System designers need to address this difference

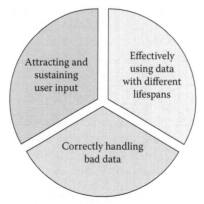

**FIGURE 10.3**
System design factors in crowd-powered approaches to developing assistive navigation aids.

in both algorithms and user interaction. Third, all crowdsource systems encounter bad data. This typically originates from user error and occasionally from bad actors. While the former is often due to inadvertent mistakes (e.g., pressing the wrong button and poor interpretation of a request), the latter is a real concern for people with disabilities and may require careful user interaction design. The presence of error requires crowd system designers to use good interaction design, error checking, and heuristics.

### 10.2.3 Localization and Constrained Path Planning

Localization in indoor environments is an attractive capability for many reasons. For blind travelers, transit stations are often complex indoor environments that are difficult to navigate. Directional remote infrared audible signs anchored to infrastructure have proved useful in terminals and as stand-ins for audible stop annunciators on buses (Miller, 2012). The main weakness for such systems is their cost (Miller, 2012). This approach is much more expensive to deploy than iBeacon, radio-frequency identification, near field communication, quick response code tag approaches (e.g., Mau et al., 2008), or smartphone localization (e.g., Sendero GPS). All of these approaches are inflexible to new information and unable to provide manipulation assistance or dynamic guidance. There are also efforts to provide information to mobile device users designed specifically for people with disabilities. For example, a number of groups have explored delivery of transit information via mobile devices (e.g., Biagioni, Agresta, Gerlich, & Eriksson, 2009), and other approaches have been used to support overall system navigation for travelers with cognitive disabilities (Repenning & Ioannidou, 2006). Some of these systems include functionality relevant to blind users, specifically automatic identification of deviations from a planned route and prompts prior to upcoming stops.

In many indoor navigation scenarios, localization is critical to obtaining relevant information or help. Accurate information allows users to rendezvous with someone who can help them and discriminates between nearby features (e.g., up escalator vs. down escalator). However, localization in unfamiliar settings is a significant challenge for blind travelers. Location-aware applications in outdoor environments can use GPS tracking to provide directions and locate places of interest. However, these systems suffer from their dependency on GPS and can therefore function accurately only when there is a clear view of the requisite number of satellites. In the absence of GPS, achieving sufficient localization accuracy on a consumer device is extremely challenging. Several GPS-free localization techniques have been studied, including inertial sensors, ultrasound, 802.15.4 radio, and others (e.g., Evennou & Marx, 2006; Otsason, Varshavsky, LaMarca, & de Lara, 2005; Priyantha, Chakraborty, & Balakrishnan, 2000). These systems have been able to demonstrate relatively high accuracy but suffer from being impractical to deploy in the mass consumer market. The inertial systems use expensive and bulky tactical-grade inertial measurement units that are not available to most. Similarly, ultrasound and 802.15.4 require installation and use of custom hardware. The challenge remains to develop a system that will reach mass consumer adoption. The mobile phone is one of the most promising avenues for a variety of tools because of its ubiquity, available sensing and processing power, and broad social acceptance.

In the future, many assumptions made in state-of-the-art indoor localization solutions will need to be relaxed, and researchers need to explore how smartphone apps, other technology solutions (such as robots and custom devices), and humans can collaborate to improve assistive localization for blind travelers (shown in Figure 10.4). Assistive localization will therefore add new requirements for interaction and cooperation. While blind travelers may need help with localization, they are rarely totally lost. They often have a significant understanding and memory of what steps they have taken and their surroundings. Effective cooperation mechanisms will therefore take advantage of their situational awareness when enabling assistive interactions in the context of localization. In addition to the situational awareness of the blind travelers, multiple complementary localization approaches can be combined using sensors on smartphones, sensors on robots operating in that environment, and ubiquitous sensors in the environment. A first obvious option is to use the onboard inertial systems on the smartphones and robots. Although reasonably accurate over the short term, this is not a complete solution because the localization error can significantly increase over time and with directional changes. Furthermore, magnetometers can be disrupted by magnetic anomalies, which are relatively common in urban environments. A second approach is Wi-Fi localization. This approach has the advantage of bounded error and the ability to localize in a global context. It works by examining the identities and signal strengths of nearby Wi-Fi access points. From these identities, the system can determine a coarse location, since it must be in an area that is in the intersection of the ranges of all the access points it can see. Comparing a map relating Wi-Fi signal

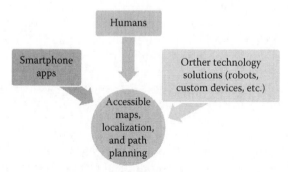

**FIGURE 10.4**
Envisioned future approach to accessible localization and route planning.

strengths to locations, a probability distribution of the current location can be obtained. Observing signal strength patterns over time, the system can converge on the location that is most likely. The primary disadvantage of this method is that it is difficult to generate such a map relating signal strengths to location, because the propagation of radio signals is hard to simulate, and measuring the data empirically is laborious. However, robots could assist with building and maintaining these maps. A third approach can use Global System of Mobile Communications (GSM), the protocol for mobile telephony. The advantage of GSM is that it has nearly 100% coverage in areas frequented by people in the United States. Like Wi-Fi, GSM can provide a coarse location estimate from the identities of nearby mobile phone towers. Since each tower covers a much larger area than a Wi-Fi access point, the estimate can be bounded to only within several miles.

A variety of modalities must also be explored to achieve localization with sufficient robustness and resolution to be effective for assisting blind travelers to localize in unfamiliar environments. Some options to be explored include crowdsourcing using photographs taken via the smartphone camera, using location information broadcast from stationary kiosks at key locations in the environment, and seeking help from other agents in the indoor environment when needed. The idea of enabling blind smartphone users to aim a camera at an object and receive help from an automated system is not new, and some approaches utilize crowdsourcing to label objects (e.g., Bigham et al., 2010). Unfortunately, this incurs a delay due to crowd response time. Bootstrapping initial crowd instruction to a more automated response is a solution of interest to the research community.

Localization alone is insufficient when navigating an unfamiliar environment. Effective route planning and path following are also important components of this task. Key to any navigation task is dynamic and efficient route planning. Route planning requires access to some form of map representation of the space to be navigated. Therefore, two critical components of any solution for assisting blind travelers must include a methodology for accessible map representation and relevant route planning. If we are to give travelers with disabilities the maximum flexibility when navigating, we need to ensure that the maps and other relevant information are accessible to them, as well as to the route-planning algorithms. While accessible maps address the interface between the traveler and the map information, they do not facilitate map representation that can be harnessed by route-planning algorithms that can be used by smartphone apps or other technology solutions. Therefore, another important aspect of related future research is determining the best forms of map representation that can capture all of the useful information, be stored efficiently on mobile devices when needed, and inform route-planning algorithms. Hierarchical maps are a likely option for representing the environment in large search spaces. The hierarchical graph allows for a compressed representation of large map details and is flexible to allow dynamic changes to be made. For example, if during travel a road is blocked because of construction or a bus is canceled, this can be annotated and subsequently represented in the map. In general, route planning requires dynamic computations of efficient or optimal paths. In route-planning applications on a mobile phone, the speed and efficiency of the algorithms can be restricted by the onboard computation capabilities. Therefore, an effective route planner should have both high throughput and low delays in terms of query processing and should be capable of dynamically replanning. Humans, other technology solutions, and smartphone apps may also need to collaborate in meaningful ways to plan relevant routes, especially if the scenario involves the evacuation of a building in an emergency.

## 10.3 Longer Term Vision

Advances in human-robot interaction, assistive robots, and team coordination lead to new possibilities in realizing the vision of safe and independent indoor navigation for visually impaired people. Furthermore, the design of smart cities must incorporate strong elements of accessibility to ensure cities of the future are truly accessible to all people, including the visually impaired. Potential directions of relevant research that should be investigated in the longer term are explored next.

### 10.3.1 Assistive Robots

The ubiquitous white cane is a tried-and-true method for local navigation. However, these have a limited perception range. The elbow method is a great option when a sighted human is available. When a blind person is walking with a sighted person, he or she will often grip the sighted person's elbow and use force and proprioception to change trajectories while walking. In fact, this can sometimes be the fastest mode of pedestrian travel for a blind person. In this respect, humanoid robots are a fascinating option for guided travel. Dogs are also an option, but a fully trained dog guide can cost as much as $42,000,* and their limited availability (Eames & Eames, 1995) results in low market penetration. There have been attempts to develop robot guide dogs (e.g., Galatas, McMurrough, Mariottini, & Makedon, 2011), but it may be difficult to justify an expensive robot for a single user. In contrast, a robot that serves a specific, busy location is not only possible but increasingly likely for malls and other high-traffic areas (e.g., Zheng, Glas, Kanda, Ishiguro, & Hagita, 2011).

Robot arms for human assistance have been available for over a decade but have generally experienced limited adoption for a variety of reasons.

---

* See Guide Dogs of America: http://www.guidedogsofamerica.org/1/mission/.

One of these is the challenge of mapping joystick and button controls to high-degree-of-freedom robot arms. This area has been an ongoing challenge because of limitations in autonomy and computer perception. Instead, recent advances have focused on providing better ways of including a human in the loop. For example, researchers at the Quality of Life Technology Center (QoLT) Engineering Research Center (ERC) at Carnegie Mellon University have employed a three-pronged approach to addressing this problem in parallel to their efforts to advance autonomous robot arm and computer vision techniques. The first model is to allow users to directly grab and move robot arms using sensors on the exterior of the robot arms (e.g., Markham & Brewer, 2009). The second approach is to employ remote assistants who can work with local users via teleoperation to complete tasks. For example, Cooper et al. (2012) demonstrated wheelchair-mounted robot arm tasks where the local user received periodic assistance from a remote helper. Finally, Dragan and Srinivasa (2012) used computer vision to translate video of the user's upper body motion into robot actions, complete with seamless handoffs to the robot for autonomous task completion. Further promising research directions in the area of assistive robots are introduced in Figure 10.5 and explored next.

There are interesting opportunities to use physical interaction with robots to both learn new tasks and convey assistance. For example, the Rethink Robotics Baxter robot is a commercial system designed for light factory duty that can be taught new physical actions and perception tasks through direct manipulation of the robot's arms. The ability to grasp and teach a robot through arm manipulation is an active research area beyond the Rethink Robotics team (e.g., Akgun, Cakmak, Yoo, & Thomaz, 2012). Eliminating the need to employ a programmer provides opportunities for local stakeholders and fellow users to create interactions specific to the environment. This harnesses the power of the crowd to develop specialized information and allows users to introduce local color. For example, a robot at Manhattan's Grand Central Station may be taught behaviors that vary greatly from one in Washington, DC's Union Station.

Direct manipulation also mitigates two key interaction barriers for people who are blind. A robot arm, or other graspable component, acts as a display

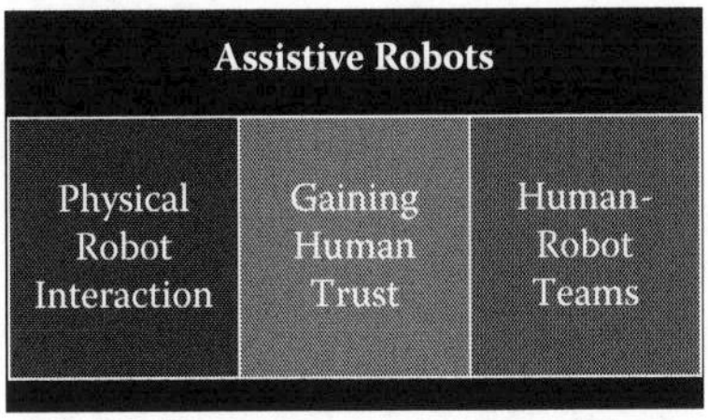

**FIGURE 10.5**
Promising research topics involving assistive robots for blind travelers.

medium in lieu of a visual interface. Second, such interaction is also inherently three dimensional and grounded. Natural language works well for many interactions involving directions (e.g., Kollar, Tellex, Roy, & Roy, 2010), but there are still interaction challenges due to mental rotation and house/stage directional confusion. A robot arm can be moved through a sequence of directions to convey a pathway through a complex environment while providing spoken directions (e.g., "Go down the left-hand hallway, up the stairs on the right, and make a U-turn."). The redundant modality will increase path memory and limit direction inversions stemming from improperly grounded perspective. This approach also supports foreign language speakers and people who process spatial information visually rather than verbally. Remote infrared audible signs, which are mostly used for labeling physical places, also provide directional information. The directionality of the infrared LED light can be focused much like a flashlight so users can capture an IR beam and follow it toward the beacon. It is easy to detect departure from the beam since guidance stops. Moving beacons have been tested as silent bus route annunciators (Golledge, Marston, & Costanzo, 1998), which suggest it may be possible to mount some type of beacon or fiducial on a robot arm for dynamic beacon positioning.

An important element of any interaction with an assistive robot, especially for systems providing assistance to blind users, is the issue of trust. Research in this area demonstrates the importance of ensuring accurate performance during early tasks (Desai et al., 2012), deferential behavior when errors are made (M. K. Lee, Kiesler, Forlizzi, Srinivasa, & Rybski, 2010), and expression of system confidence changes (Desai, Kaniarasu, Medvedev, Steinfeld, & Yanco, 2013). Furthermore, low reliability in robot autonomy early in navigational tasks can lead to significantly reduced trust and poorer utilization of autonomy when compared with low reliability that occurs later in the experience (Desai et al., 2012, 2013). These findings imply that learning strategies for robots should require a high bar prior to transition to autonomy and that emphasis should be placed on assistance tasks that occur early in a participant's exposure to the robot. For example, a guide robot should demonstrate good performance in the immediate vicinity of the user's first encounter. This also aligns with findings that show that past use of automated systems increases likelihood of continued use, even when operators think they might be able to perform the task better than an automated component (J. D. Lee & Moray, 1994). Finally, the issue of accurate and appropriate speech interaction with robots is also relevant to envisioned types of assistive navigation technology (e.g., M. K. Lee et al., 2010).

Assistive human-robot interactions do not need to be limited to interactions between a single human and a single robot. The scenario of human-robot teams engaging in assistive tasks is another powerful research direction that merits investigation. While many approaches have been designed to accommodate different human-robot team coordination requirements, assistive navigation scenarios necessitate coordinating teams in more unusual settings where humans and robots must act as peers (Dias et al., 2008). Within the topic of human-robot teams, there are interesting research opportunities for teams where the composition of the team is not previously known or fixed and where members joining the team can vary in their capabilities, expertise, and knowledge of the task. This is a new area of research where relatively little work has been done in the past. Gaining and maintaining situational awareness is one of the biggest challenges in such team settings. Situational awareness is a key factor in executing early and successful interventions and in making decisions for seeking additional assistance. In teams with multiple mobile agents, it is not sufficient to capture information in a single user interface, and customization of the state information for the different members of the team may be required. Furthermore, the state of the humans and the dialogue and gestures that are a natural part of human-to-human communication must be captured and made transparent to the robots on the team since interventions might be carried out by robots. The robots themselves may be heterogeneous and require different modalities of information representation. Situation awareness in dynamically formed "pickup" teams is also difficult because we must be able to accommodate new capabilities and resources as members join the team, and we must be able to expose the state of the current team to the new member quickly and effectively. Several research efforts have focused on a variety of communication strategies for human-robot teams that include tools such as graphical user interfaces, 3D interactive environments, dialogue, and gestures (Driewer, Sauer, & Schilling, 2007). However, there is still much to be accomplished in this area of research, especially in the context of including people with disabilities on these teams.

### 10.3.2 Information Exchange and Object Manipulation

Exchanging information and manipulating objects of interest are the foundations for many interactions. In the context of blind travelers, robots can assist with providing a variety of travel-related information, assist humans with locating lost objects such as a mobile phone or a dropped white cane, help blind travelers sort unfamiliar currency, and more. A key component of these tasks will be accessible interfaces that allow blind travelers to effectively communicate their intents, needs, and state to the robots and vice versa. A variety of interface options must be explored to enable such exchanges.

These interfaces will enable communication that can lead to the exploration of several interaction modalities. Several of these interaction modes that could enhance navigation for blind people are shown in Figure 10.6 and described next.

For example, robot manipulators can assist with sorting or transferring objects of interest, while mobile robots can assist with searching an area for a person of interest or reading a sign or label of interest to a blind traveler. The interactions could also include a teaching or learning component where

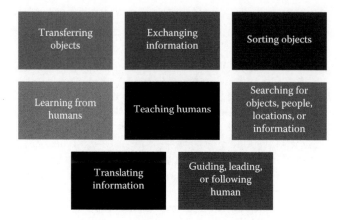

**FIGURE 10.6**
Potential interactions between assistive robots and blind travelers.

the blind traveler must teach a robot how to perform a specific task such as how to identify a specific object of interest or how to recognize a gesture that indicates the robot should slow down if leading the blind traveler. Robots could also teach a blind traveler new skills such as how to identify specific coins in the local currency, a specific sound signature that alerts the blind traveler that a location of interest is nearby, or a specific gesture that alerts the visually impaired human to potentially dangerous situations.

Beyond one-on-one interactions, there are opportunities for cooperation when exchanging information or manipulating objects. For example, robots with different capabilities might team up to locate and retrieve an object of interest to be transferred to a blind traveler. Similarly, robots can transfer some knowledge and skills to other robots or share information that enables the execution of complex tasks. Finally, robots can assist blind travelers to recruit sighted humans nearby to help in difficult situations, or humans (both sighted and visually impaired) might team up with robots to manipulate a larger object or coordinate to retrieve information for a complex query.

### 10.3.3 Toward Smart Accessible Cities

As mentioned earlier, there is a strong need for integrated location, travel, and points of interest information (Geo-Access Challenge Team, 2011). The smart cities vision advocated by many stakeholders will directly impact the ability to collect, collate, and disseminate such information, thus leading to easier trip planning, route execution, and general navigation. An example of this future was illustrated in IBM's Access My NYC demonstration.* This prototype integrated public and private transportation options, points of interest, social media, and accessibility search filters in a mobile Web application.

---

* See http://www-03.ibm.com/able/accessibility_research_projects/AccessMyCity.html.

Similar systems are likely in the near future because of numerous government initiatives focused on providing open data to support innovative applications by third-party developers. Another major US government initiative that is likely to enable new functionalities is the possible federally mandated deployment of dedicated short-range communications (DSRC) for transportation safety applications. Aside from generally improved transportation safety and efficiency, DSRC will provide significant benefit to people who are blind and low vision. First, this will allow similar functionality to earlier niche systems that labeled transit buses, intersections, and walk signals with remote infrared audible signs and other wireless communications (Morton & Yousuf, 2011). Second, DSRC has the potential to improve localization in dense urban environments.

Navigating an environment under standard conditions entails different constraints and requires different strategies in comparison to navigating the same environment in an emergency evacuation scenario. As shown in Figure 10.7, assistive technology tools have to be location-aware, situation-aware, and user-aware in order to succeed in these extreme situations. Location-awareness is needed to successfully guide the user from the current location to safety. Situation-awareness is needed to recognize the emergency and employ the most suitable strategies in seeking help and assisting the user to reach a safe location. Finally, user-awareness is needed to understand the user's capabilities, needs, and preferences and execute the safety plan most suited to the user. The design and implementation of smart cities can take into account these needs and incorporate relevant technology solutions to ensure emergency evacuation plans are also accessible to visually impaired people.

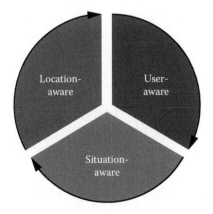

**FIGURE 10.7**
Three key requirements for assistive technology tools in emergencies.

## 10.4 Summary

Indoor wayfinding and navigation technology to assist blind users is available today with limited performance, and research groups around the world are continuing to improve the state of the art in this area. This field is still in its nascent stages, and many navigation needs of blind commuters have yet to be addressed in meaningful and effective ways. This chapter explored some of the possible pathways researchers and innovators can follow to further enhance and facilitate the indoor navigation experience for visually impaired travelers.

Although public transportation options in many cities have enabled blind travelers to more independently get from point A to point B, navigating through complex spaces such as transit stations can be daunting for most visually impaired people. Localization and wayfinding are two of the biggest hurdles that need to be alleviated in order to create more accessible indoor spaces. In addition, solutions ought to be responsive to the dynamic nature of these environments so that travelers can adapt to real-time changes in their locale. User interfaces for assistive devices must be improved to provide clearer audio feedback, more landmarks, enhanced customizable instructions, and other key features that can help render assistive tools more effective and efficient. Limitations of existing localization techniques can be circumvented by incorporating alternate sensors and crowdsourcing into the localization framework. Furthermore, more effective representations of large-scale floor plans and maps, as well as improved path-planning algorithms, are additional important areas of focus for the future of assistive navigation technology.

Assistive navigation systems that are based on crowdsourcing have proved to be somewhat successful in offering commuters more up-to-date information on travel considerations. However, two key challenges manifest with such systems: (1) obtaining a sufficiently large "crowd" to contribute information and (2) ensuring that information received is accurate and reliable. Smartphones have emerged as a very useful tool for delivering information to visually impaired travelers. Still, the interface for such devices could be enhanced to facilitate users' interaction with data and also to incorporate more effective directional instructions, particularly when white canes and guide dogs are insufficient to assist them.

Navigation solutions supported through human-robot interactions could also serve an important role in the future of assistive technology. Moving toward robotic technology that can be directly manipulated by users could remove the overhead of customized programming to match specific use-case scenarios. For assistive robots to become a reality, the situational awareness of robots will need to be accurate and detailed enough to offer the necessary guidance to blind travelers and also to establish these robot guides as trustworthy navigation aids. Much work can also be done to promote trust in robot guides as assistive navigation tools for blind travelers. To this end, failure rates at least for base-level functions need to be minimized, and input from the blind community should be incorporated into the design, deployment, and development of these technologies.

In the future, we can anticipate that robotic and other technologies will be integrated into the navigation experience such that these tools can handle real-time information exchange with users and also manipulate physical objects to assist in various aspects of the navigation process (e.g., open doors, locate and pick up dropped items). The grander vision for accessible navigation solutions will certainly involve more systematic improvements made to the general structure of urban indoor environments. Furthermore, in the evolution of robotics navigation technology, specific attention will need to be dedicated to ensuring that robots are capable of interacting with visually impaired humans. Carving out meaningful human-robot interaction methodologies is a crucial step in this process of developing assistive robots for blind travelers. Moreover, smartphones and other mobile devices will need to be equipped with more responsive and accessible interfaces so that they can serve as effective modalities for information access and exchange with blind travelers. All these considerations will need to be incorporated into the notion of smarter cities of the future, which should facilitate indoor navigation for the visually impaired.

Finally, assistive navigation technology will have to tackle the complicated issues pertaining to emergency evacuation scenarios involving visually impaired travelers. In such circumstances, blind individuals will need to receive sufficient guidance to safely and quickly exit a building while confronting a more chaotic indoor atmosphere. Situations requiring emergency evacuation pose a significant threat to the safety of blind travelers, since existing infrastructure and tools do not sufficiently support their needs in such an event. This is a critical shortcoming that needs to be addressed in the future of assistive navigation technology development.

## Acknowledgments

Preparation of this chapter was supported by the National Science Foundation (NSF-NRI Award Number 1317989), the US Department of Transportation University Transportation Centers Program, and the National Institute on Disability and Rehabilitation Research under the RERC-APT (Grant Numbers H133E130004 and H133E080019). Any opinions, findings, conclusions, and recommendations expressed in this work are those of the authors and do not necessarily reflect the views of the sponsors. The authors also thank the many members of the TechBridgeWorld research team at Carnegie Mellon University who provided assistance and insight that enhanced this work in many ways, notably, Ermine Teves, Sarah Belousov, Dr. Balajee Kannan, and M. Freddie Dias.

# References

Akgun, B., Cakmak, M., Yoo, J. W., & Thomaz, A. L. (2012). Trajectories and keyframes for kinesthetic teaching: A human-robot interaction perspective. In *Proceedings of the ACM/IEEE international conference on human-robot interaction* (pp. 391–398). New York, NY: IEEE.

Biagioni, J., Agresta, A., Gerlich, T., & Eriksson, J. (2009). TransitGenie: A context-aware, real-time transit navigator. In *Proceedings of the 7th ACM conference on embedded networked sensor systems (SenSys)* (pp. 329–330). New York, NY: ACM.

Bigham, J. P., Jayant, C., Ji, H., Little, G., Miller, A., Miller, R. C., Miller, R., Tatarowicz, A., White, B., White, S., & Yeh, T. (2010). Vizwiz: Nearly real-time answers to visual questions. In *UIST '10 proceedings of the 23rd annual ACM symposium on user interface software and technology* (pp. 333–342). New York, NY: ACM.

Cooper, R. A., Grindle, G. G., Vazquez, J. J., Xu, J., Wang, H., Candiotti, J., Chung, C., Salatin, B., Houston, E., Kelleher, A., Cooper, R., Teodorski, E., & Beach, S. (2012). Personal mobility and manipulation appliance: Design, development, and initial testing. *Proceedings of the IEEE, 100*(8), 2505–2511.

Dale, K., & Deutchman, S. (2010, July 20). The power of technology, the power of equality [Web blog post]. Retrieved from www.whitehouse.gov/blog/2010/07/20/power-technology-power-equality

Desai, M., Kaniarasu, P., Medvedev, M., Steinfeld, A., & Yanco, H. (2013). Impact of robot failures and feedback on real-time trust. In *Proceedings of ACM/IEEE international conference on human-robot interaction*. New York, NY: IEEE.

Desai, M., Medvedev, M., Vázquez, M., McSheehy, S., Gadea-Omelchenko, S., Bruggeman, C., Steinfeld, A., & Yanco, H. (2012). Effects of changing reliability on trust of robot systems. In *Proceedings of the ACM/IEEE international conference on human-robot interaction* (pp. 73–80). New York, NY: IEEE.

Dias, M. B., Kannan, B., Browning, B., Jones, E., Argall, B., Dias, M. F., Zinck, M. B., Veloso, M., & Stentz, A. (2008). Sliding autonomy for peer-to-peer human-robot teams. In W. Burgard, R. Dillmann, C. Plagemann, & N. Vahrenkamp (Eds.), *Intelligent autonomous systems 10.* Amsterdam, the Netherlands: IOS Press.

Dragan, A., & Srinivasa, S. (2012). Online customization of teleoperation interfaces. In *IEEE international symposium on robot and human interactive communication (RO-MAN)* (pp. 919–924). New York, NY: IEEE.

Driewer, F., Sauer, M., & Schilling, K. (2007). Discussion of challenges for user interfaces in human-robot teams. In *Proceedings of the third European conference on mobile robots*. Retrieved from http://citeseerx.ist.psu.edu/viewdoc/summary?doi = 10.1.1.84.1755

Eames, E., & Eames, T. (1995). Demographics update: Alternate estimate of the number of guide dog users. *Journal of Visual Impairment and Blindness, 89*(2), 4–6.

Evennou, F., & Marx, F. (2006). Advanced integration of WiFi and inertial navigation systems for indoor mobile positioning. *EURASIP Journal on Applied Signal Processing, 2006,* 1–11.

Galatas, G., McMurrough, C., Mariottini, G. L., & Makedon, F. (2011). eyeDog: An assistive-guide robot for the visually impaired. In *Proceedings of the 4th international conference on PErvasive Technologies Related to Assistive Environments (PETRA)* (Article 58). New York, NY: ACM.

Gaunet, F., & Briffault, X. (2005). Exploring the functional specifications of a localized wayfinding verbal aid for blind pedestrians: Simple and structured urban areas. *Human-Computer Interaction, 20*(3), 267–314.

Geo-Access Challenge Team. (2011). *Data-enabled travel: How geo-data can support inclusive transportation, tourism, and navigation through communities.* Retrieved from http://geoaccess.org/content/report-data-enabled-travel

Golledge, R. G., Marston, J. R., & Costanzo, C. M. (1998). *Assistive devices and services for the disabled: Auditory signage and the accessible city for blind or vision impaired travelers* (Working paper UCB-ITS-PWP-98-18). Berkeley: PATH Program, Institute for Transportation Studies, University of California, Berkeley. Retrieved from http://www.path.berkeley.edu/PATH/Publications/PDF/PWP/98/PWP-98-18.pdf

Jayant, C., Acuario, C., Johnson, W., Hollier, J., & Ladner, R. E. (2010). VBraille: Haptic braille perception using a touch-screen and vibration on mobile phones. In *Proceedings of the 12th international ACM SIGACCESS conference on computers and accessibility* (pp. 295–296). New York, NY: ACM.

Kane, S. K., Bigham, J. P., & Wobbrock, J. O. (2008). Slide rule: Making mobile touch screens accessible to blind people using multi-touch interaction techniques. In *Proceedings of the 10th international ACM SIGACCESS conference on computers and accessibility* (pp. 73–80). New York, NY: ACM.

Kane, S., Morris, M., Perkins, A., Wigdor, D., Ladner, R., & Wobbrock, J. (2011). Access overlays: Improving non-visual access to large touch screens for blind users. In *Proceedings of the 24th annual ACM symposium on user interface software and technology* (pp. 273–282). New York, NY: ACM.

Kannan, B., Kothari, N., Gnegy, C., Gedaway, H., Dias, M. F., & Dias, M. B. (2014). Localization, route planning, and smartphone interface for indoor navigation. In A. Koubaa & A. Khelil (Eds.), *Cooperative robots and sensor networks* (pp. 39–59). Berlin, Heidelberg, Germany: Springer.

Kehret, G., Miele, J., & Landau, S. (2011). *Development of smartpen-based audio/tactile transit station maps for travel planning and wayfinding.* Paper presented at the CSUN Technology and Persons with Disabilities Conference.

Kim, J.-K., & Zatorre, R. J. (2010). Can you hear shapes you touch? *Experimental Brain Research, 202*(4), 747–754.

Klatzky, R., Marston, J., Giudice, N., Golledge, R., & Loomis, J. (2006). Cognitive load of navigating without vision when guided by virtual sound versus spatial language. *Journal of Experimental Psychology: Applied, 12*(4), 223–232.

Kollar, T., Tellex, S., Roy, D., & Roy, N. (2010). Toward understanding natural language directions. In *Proceedings of the ACM/IEEE international conference on human-robot interaction* (pp. 259–266). New York, NY: IEEE.

Lee, J. D., & Moray, N. (1994). Trust, self-confidence, and operators' adaptation to automation. *International Journal of Human-Computer Studies, 40*(1), 153–184.

Lee, M. K., Kiesler, S., Forlizzi, J., Srinivasa, S., & Rybski, P. (2010). Gracefully mitigating breakdowns in robotic services. In *Proceedings of the ACM/IEEE international conference on human-robot interaction* (pp. 203–210). New York, NY: IEEE.

Markham, H. C., & Brewer, B. R. (2009). Development of a skin for intuitive interaction with an assistive robot. In *Engineering in Medicine and Biology Society, 2009. EMBC 2009. Annual international conference of the IEEE* (pp. 5969–5972). New York, NY: IEEE.

Marston, J. R., Loomis, J. M., Klatzky, R. L., & Golledge, R. G. (2007). Nonvisual route following with guidance from a simple haptic or auditory display. *Journal of Visual Impairment and Blindness, 101*(4), 203–211.

Marston, J. R., Loomis, J. M., Klatzky, R. L., Golledge, R. G., & Smith, E. L. (2006). Evaluation of spatial displays for navigation without sight. *ACM Transactions on Applied Perception, 3*(2), 110–124.

Mau, S., Melchior, N., Makatchev, M., & Steinfeld, A. (2008). *BlindAid: An electronic travel aid for the blind* (CMU-RI-TR-07-39). Pittsburgh, PA: Robotics Institute, Carnegie Mellon University. Retrieved from http://www.cs.cmu.edu/~mmakatch/papers/blindaid.pdf

Miller, M. (2012). *Remote infrared audible signage pilot program* (FTA Report No. 0012). Washington, DC: Federal Transit Administration, U.S. Department of Transportation. Retrieved from http://www.fta.dot.gov/documents/FTA0012_Research_Report_Summary.pdf

Morton, T., & Yousuf, M. (2011). *Technological innovations in transportation for people with disabilities workshop summary report* (FHWA-HRT-11-041). Washington, DC: Office of Operations Research and Development, Federal Highway Administration, U.S. Department of Transportation. Retrieved from http://www.fhwa.dot.gov/advancedresearch/pubs/11041/

Otsason, V., Varshavsky, A., LaMarca, A., & de Lara, E. (2005). Accurate GSM indoor localization. In *Proceedings of the international conference on ubiquitous computing (Ubicomp)* (pp. 141–158). Berlin, Heidelberg, Germany: Springer-Verlag.

Priyantha, N. B., Chakraborty, A., & Balakrishnan, H. (2000). The cricket location-support system. In *Proceedings of the 6th annual ACM international conference on mobile computing and networking (MOBICOM)* (pp. 32–43). New York, NY: ACM.

Repenning, A., & Ioannidou, A. (2006). Mobility agents: Guiding and tracking public transportation users. In *Proceedings of the working conference on advanced visual interfaces (AVI)* (pp. 127–134). New York, NY: ACM.

Rice, M., Jacobson, R. D., Golledge, R. G., & Jones, D. (2005). Design considerations for haptic and auditory map interfaces. *Cartography and Geographic Information Science, 32*(4), 381–391.

Steinfeld, A., Zimmerman, J., Tomasic, A., Yoo, A., & Aziz, R. (2011). Mobile transit information from universal design and crowdsourcing. *Transportation Research Record: Journal of the Transportation Research Board, 2217*, 95–102.

Su, J., Rosenzweig, A., Goel, A., de Lara, E., & Truong, K. N. (2010). Timbremap: Enabling the visually-impaired to use maps on touch-enabled devices. In *Proceedings of the 12th international conference on human computer interaction with mobile devices and services (MobileHCI)* (pp. 17–26). New York, NY: ACM.

Talbot, M., & Cowan, W. (2009). On the audio representation of distance for blind users. In *Proceedings of the SIGCHI conference on human factors in computing systems* (pp. 1839–1848). New York, NY: ACM.

Tissot, N. (2003). *Indoor navigation for visually impaired people: The navigation layer.* Zurich, Switzerland: Department of Computer Science, Institute for Pervasive Computing, ETH Zurich.

Trace Center. (2000, Spring). *Making information/transaction machines (ITMs) accessible: Final report.* Madison: Trace Research and Development Center, University of Wisconsin–Madison, U.S. Access Board. Retrieved from http://trace.wisc.edu/world/kiosks/itms/index.html

Wall, S., & Brewster, S. (2006). Feeling what you hear: Tactile feedback for navigation of audio graphs. In *Proceedings of the SIGCHI conference on human factors in computing systems* (pp. 1123–1132). New York, NY: ACM.

Way, T., & Barner, K. (1997). Automatic visual to tactile translation—Part 1: Human factors, access methods and image manipulation. *IEEE Transactions on Rehabilitation Engineering, 5,* 81–94. Retrieved from http://www.ncbi.nlm.nih.gov/pubmed/9086389

World Health Organization. (2012). *Disability and health fact sheet N°352.* Retrieved from http://www.who.int/mediacentre/factsheets/fs352/en/index.html

Zheng, K., Glas, D. F., Kanda, T., Ishiguro, H., & Hagita, N. (2011). How many social robots can one operator control? In *Proceedings of the ACM/IEEE international conference on human-robot interaction* (pp. 379–386). New York, NY: IEEE.

# 11

## Protecting Privacy in Indoor Positioning Systems

Balaji Palanisamy

James Joshi

### CONTENTS

*Abstract:* An indoor positioning system is a network of devices used to wirelessly locate objects or people inside a building. When it comes to the widespread adoption of indoor positioning and navigation tools among users, technology is not often a barrier, as recent advancements in positioning technologies have produced a number of low-cost accurate embedded positioning devices. However, privacy concerns remain the major consumer pushback for a wider acceptance of such services among users. An indoor navigation service enables smartphone users to navigate indoors and share indoor location-based information. It thus generates a huge amount of potentially sensitive information, including the activities of smartphone users' movements, information searched by them, and the content they share with their neighboring peers. Even though it is believed that smartphone users are aware that such positioning services are tracking their mobile activities, a large fraction of the smartphone user populations are concerned about their location and other sensitive data generated through such services. Overcoming these concerns is a top priority for both service vendors who provide indoor location technologies to retailers and mobile consumers who are conscious of their privacy when they navigate indoors. In this chapter, we introduce the privacy risks and threats related to indoor positioning systems and the state-of-the-art countermeasures to protect against them. We also discuss the new challenges in developing privacy-enabling indoor positioning and navigation tools and their merits and demerits.

## 11.1 Introduction

Advances in sensing and positioning technology, fueled by wide deployment of wireless local area networks, have made many devices location-aware. An indoor positioning system is a network of devices used to wirelessly locate objects or people inside a building. In the past, people have widely relied on the global positioning system (GPS) for location information, which, however, does not work well in indoor spaces or urban canyons with streets cutting through dense blocks of high-rise buildings and structures. GPS technology often requires a clear view to communicate with satellites and does not work well when its signals become attenuated or scattered by roofs, walls, and other objects.* Also, GPS technology is only one third as accurate in the vertical direction as it is in the horizontal, which makes it impossible to locate a person or an object in the floors of skyscrapers. In contrast, location-based services (LBSs) for indoor positioning systems commonly provide position based on Wi-Fi, cellular connectivity, ultrawideband, or radio-frequency identification. Recent mobile location-based applications have utilized such positioning technologies to provide clients with location information services.

The collection and transfer of location information about a particular subject can have important privacy implications. Concrete examples of indoor LBSs include searching nearest points of interest (e.g., "Where is the nearest coffee shop in the mall?"), spatial alerts (e.g., "Remind me when I am walking close to the ATM"), and location-based social networking (e.g., "Where is my friend, Tom, in the mall?"). Indoor navigation systems can significantly help firefighters (and other emergency responders) entering an unfamiliar building in low-visibility situations and enable others to know where they are in 3D space. Such location-based and navigation services would require the service provider to track the location information of their mobile users in order to deliver them the required service. In general, indoor location-based tools can require either a snapshot or a continuous exposure of location information. Snapshot queries, such as obtaining the floor map of the nearest coffee shop in a mall, require only a snapshot exposure of the location to obtain the information. On the other hand, certain indoor location-based applications can require continuous location exposure such as finding dynamic directions for navigation to a particular store inside a mall. Although indoor LBSs

---

* See http://www.sciencedaily.com/releases/2012/12/121217140629.htm.

offer many interesting and life-enhancing experiences, they certainly open doors for new security risks that can endanger the location privacy of mobile clients ("Authorities," 2002; Karger & Frankel, 1995).

Privacy risks in indoor LBSs arise out of two main sources of information: (1) the massive amount of location traces generated by indoor positioning devices exposed to the untrusted LBS provider can result in serious location privacy threats to the user, and (2) the content accessed by the mobile users can itself cause serious violation of user privacy.

*Location privacy* is a particular type of information privacy. According to Beresford and Stajano (2003), location privacy is defined as the ability to prevent other unauthorized parties from learning one's current or past location. Location privacy threats refer to the risks that an adversary can obtain unauthorized access to raw location data or derived or computed location information by locating a transmitting device, hijacking the location transmission channel, and identifying the person using a mobile device. For example, location information can be used to spam users with unwanted advertisements or draw unwanted inferences from victims' visits to clinics, doctors' offices, entertainment districts, religious activities, or political events. In extreme cases, unauthorized disclosure of private location information can lead to physical harm, for example, in stalking or domestic abuse scenarios. Even though some LBS providers (such as Google Maps) have a well-stated privacy policy statement, such a privacy statement is primarily for not exposing the collected information to public and commercial uses. Thus, there are still inherent risks in continuous collection of location information by the LBS provider, as there are channels of attacks beyond the control of the LBS provider and the protection of the privacy policy statement, including insider attacks. For instance, there was a recent incident in Google where a Google engineer spied on four underage teens for months before the company was notified of the abuses (Chen, 2010).

*Content privacy* refers to the ability of the users to obtain location-based information from untrusted service providers in such a way that the unauthorized parties do not learn about the content that is being requested by the mobile clients. For instance, a user may request a query to locate a particular product in a supermarket but may not wish the untrusted service provider to know that the user is interested in that particular product. Content privacy is an equally important issue and is often complementary to location privacy.

In the past, a fair amount of research efforts have been dedicated to protecting location and content privacy risks of mobile travelers. In this chapter, we define and introduce the various location privacy and content privacy attacks that can happen in indoor navigation systems and their associated risks. We then discuss various common location and content privacy protection mechanisms and study their applicability in the context of indoor navigation systems. We discuss the technical challenges in applying such protection models for indoor LBSs and the possible impact on user service quality and privacy levels. We outline some future research directions along our discussions and discuss the relative merits of various solutions.

---

## 11.2 Privacy Risks and Attacks

In general, an adversary is assumed to be associated with an untrusted LBS provider who may obtain a time series of cloaked locations or locations with pseudonyms used by the mobile clients. The adversary is considered successful if he or she can infer the correct linkage of a user's real identity to his or her location. Thus, the overall goal of the adversary is to track the whereabouts of users by tracking the locations of the users and the contents requested by them in their service requests. An adversary can utilize such location information to infer details about the private life of individuals such as their political affiliations, alternative lifestyles, or medical problems or the private businesses of an organization such as new business initiatives and partnerships. When users use a pseudo-identity (pseudonym) instead of their real identity, once the attacker knows the real identity behind the pseudonym, any queries asked by the user with the pseudonym can be linked to the original identity, which is a violation of the user's privacy. Thus, the location privacy attacks can be broadly classified as follows (Shokri, Theodorakopoulos, Le Boudec, & Hubaux, 2011):

- *Tracking attack:* If the adversary's objective is to find out the whole sequence (or a partial subsequence) of the events in a user's trace, the attack is called a tracking attack. An example of a very general tracking attack is the one that aims to recover the actual trace of each user. That is, it targets the whole set of users and the whole set of time instances, and it asks for the most likely trace of each user or even for the whole probability distribution of traces for each user.

- *Localization attack:* The attacks that target a single event (at a given time instant) in a user's trace are called localization attacks. This attack is an example of presence-absence disclosure attacks: they infer the relation between users and regions over time.

- *Meeting disclosure attack:* In contrast, if the physical proximity between users is of the adversary's interest, we call the attack a meeting disclosure attack (i.e., who meets whom possibly at a given place and time).

- *Content privacy attack:* Unlike the above-mentioned attacks that focus on inferring the location of a user, content privacy attacks focus on inferring the location-based content requested by the user.

By understanding the information requested by the mobile clients, an attacker could invade his or her privacy by inferring the user's personal interests.

In the past, a number of studies have addressed location and content privacy concerns in the context of LBSs, and their techniques can be broadly classified into two categories, namely, (1) techniques that limit the granularity of location information through spatial or temporal cloaking (Bamba, Liu, Pesti, & Wang, 2008; Gedik & Liu, 2005; Gruteser & Grunwald, 2003; Mokbel, Chow, & Aref, 2006; Wang & Liu, 2009) and (2) techniques that break the continuity of location exposure through the use of techniques such as mix-zones (Beresford & Stajano, 2003; Buttyan, Holczer, & Vajda, 2007; Freudiger, Raya, Flegyhazi, Papadimitratos, & Hubaux, 2007; Freudiger, Shokri, & Hubaux, 2009; Palanisamy & Liu, 2011). Spatial cloaking approaches hide the exact point location of the mobile objects using a larger spatial region that reduces the attacker's certainty in inferring the exact location of the mobile client. More specifically, the spatially cloaked region is constructed to ensure that at least $k$ users (*location k anonymity*) are located in it and contains $l$ different static locations (*location l diversity*). On the other hand, temporal cloaking hides the exact time instance when the mobile user is located at a particular location by delaying the query by a randomly chosen time delay during which at least $k$ mobile users visit the location, and therefore the query could have originated from any of the $k$ users. While spatial and temporal cloaking techniques mask the spatial and temporal granularity of location exposure, mix-zone-based techniques break the continuity of location exposure by changing pseudonyms inside mix-zones where no applications can track user movements. Mix-zones are regions in space where a set of users enter and change pseudonyms in such a way that the mapping between their old and new pseudonyms is not revealed. Mix-zones anonymize user identities and break the continuity of user trajectories. We discuss these techniques and their merits and demerits for indoor navigation applications in the next sections.

## 11.3 Spatial-Cloaking-Based Techniques and Challenges

The first category of location privacy protection schemes is represented by location cloaking techniques (Bamba et al., 2008; Gedik & Liu, 2005; Gruteser & Grunwald, 2003; Mokbel et al., 2006; Wang & Liu, 2009). In this section, we discuss the principles behind spatial cloaking and the challenges in adopting this model to indoor LBSs. Spatial location cloaking typically adds uncertainty to the location information exposed to the location query services by increasing the spatial resolution of a mobile user's locations while meeting location $k$-anonymity and/or location $l$-diversity (Bamba et al., 2008). More specifically, the spatially cloaked region is constructed to ensure that at least $k$ users (*location k anonymity*) are located in the same region, which contains $l$ different static-sensitive objects (locations). Typically, spatial cloaking techniques work well with the class of location-based applications that accept pseudonyms, such as applications that do not require the true identity of the users. In the context of LBSs and mobile users, location $k$-anonymity refers to $k$-anonymous usage of location information. A user is considered location $k$-anonymous if and only if the location information sent from the mobile user to an LBS is indistinguishable from the location information of at least $k$–1 other users. Location $l$-diversity is introduced to strengthen the privacy protection of location $k$-anonymity in situations where location information shared by the $k$ users is sensitive. Increasing $l$ value to 2 or higher significantly reduces the probability of linking a static location or a symbolic address (such as a specific store, the food court, or the restroom within a huge mall) to a mobile user. Location perturbation is an effective technique for implementing personalized location $k$-anonymity and location $l$-diversity. Cloaking methods typically perturb the location information by reducing its resolution in terms of space and time, referred to as spatial cloaking and temporal cloaking.

The concept of location $k$-anonymity was originally introduced in Aggarwal (2005), where $k$ was set to be uniform for all users. The concept of personalized location $k$-anonymity with customizable Quality of Service (QoS) specifications, first introduced in Gedik and Liu (2005), has been adopted by several others (Bamba et al., 2008; Mokbel et al., 2006). The most popular solutions for location privacy (Bamba et al., 2008; Gedik & Liu, 2005; Mokbel et al., 2006) have adopted the trusted third-party anonymization model, which has been successfully deployed in other areas such as Web browsing.[*] Two representative approaches to personalized location anonymization are the CliqueCloak algorithm, introduced in Gedik and Liu (2005), and the Casper system (Mokbel et al., 2006). The CliqueCloak algorithm relies on the ability to locate a clique in a graph to perform location cloaking, which is expensive and shows poor performance for large $k$. The Casper approach performs the location anonymization using the quadtree-based pyramid data structure, allowing fast cloaking. However, because of the coarse resolution of the pyramid structure and lack of mechanisms to ensure QoS and constrain the size of the cloaking region, the cloaking areas in Casper are much larger than necessary, leading to poor QoS perceived by the users. The PrivacyGrid approach (Bamba et al., 2008) outperforms Casper and other existing location anonymization approaches in terms of efficiency and effectiveness, producing cloaking regions that meet both location privacy and location service quality requirements.

Temporal cloaking refers to delayed cloaking of the messages by introducing bounded delay by providing a user-specified maximum temporal

---

[*] See http://www.anonymizer.com.

tolerance constraint. Although spatial cloaking algorithms in general are more scalable and have a much higher anonymization success rate, at times a few messages get dropped because of low user-density areas. This happens especially when the requested spatial resolution is too low. For example, it may happen that the current location parameters of objects are not suitable to satisfy a particular cloaking request. In such cases, the user may specify a temporal tolerance *dt* as a part of his or her privacy profile. This permits the location anonymization framework to delay the anonymization request by a maximum period of *dt* seconds; within this period, object motion patterns may change, aiding the anonymization process. In other scenarios, trade-offs will be involved between spatial accuracy and temporal latency provided by the cloaking algorithm, as delaying cloaking requests may help the system cloak messages to a smaller region.

However, in the location cloaking approach, the use of spatially cloaked resolution instead of the exact position of users does not prevent continuous exposure of location information and thus may lead to breaches of location privacy due to statistics-based inference attacks (Krumm, 2007). Another major disadvantage of location cloaking techniques is that they are not suitable for continuous queries, as they are subject to query-correlation attacks (C.-Y. Chow & Mokbel, 2007). In recent years, there have been research efforts that dealt with location privacy risks of continuous queries while using location cloaking. C.-Y. Chow and Mokbel (2007) proposed spatial cloaking using the memorization property for continuous queries. This was further used in Pan, Meng, and Xu (2009) for clustering queries with similar mobility patterns. However, this type of technique leads to large cloaking boxes, resulting in higher query processing costs, as users may not always move together. Therefore, spatial cloaking is effective only for snapshot queries but vulnerable to continuous query (CQ) attacks.

*Indoor navigation system challenges:* While spatial cloaking and temporal cloaking techniques are in general efficient approaches for protecting snapshot location-based queries, applying them for indoor location privacy brings a set of new challenges. First, user density in indoor areas is significantly higher, with many stores and rooms located in closer vicinities. A small region in indoor space has an enormous amount of location-based information, and hence it is harder for the cloaking approach to simultaneously yield both location obfuscation and highly accurate results to location-based queries. For instance, if a cloaking region that consists of location 5-diversity (e.g., consisting of the spatial regions of five stores in a shopping mall) is exposed to the LBS provider, the results of the query would be accurate only with respect to the spatial region consisting of five different stores. Hence, the quality of information provided by such anonymous services could be unacceptable. Here, we note that even though some client-side anonymous query processors can filter the less accurate results and improve the quality of the obtained information (Mokbel et al., 2006), such techniques would impose a significantly higher processing cost at

the clients or anonymizers compared to conventional location-based queries. Second, the concept of temporal cloaking may be less acceptable in an
indoor scenario, where people actually issue a query while they are navigating inside a building or browsing inside a store. Therefore, the challenge
here is to devise indoor location cloaking schemes that can yield high-quality location-based information while achieving higher *l*-diversity.

## 11.4  Mix-Zone-Based Protection and Limitations

The next category of location privacy research is embodied by mix-zone-
based approaches. In contrast to controlling the resolutions of locations used
in spatial cloaking-based location privacy solutions, mix-zones protect location privacy by changing pseudonyms at selective locations such that it is
very hard to link new pseudonyms with old pseudonyms. Mix-zone-based
techniques anonymize user identity by restricting when and where the exposure of users' positions are allowed (Beresford & Stajano, 2003). *Mix-zones* are
regions in space where no applications can trace user movements. Mix-zones
enable users to change pseudonyms such that the linking between the old
pseudonym and the new one is not revealed. The idea behind using pseudonyms instead of real identities is to disassociate the exposure of location
information from the actual identity of the person. However, when a pseudonym is used by a user for a continued duration of time, the adversary can
link a pseudonym to the user's actual identity through the inference of the
user's personal locations such as home address, office location, and other
known favorite locations. For instance, if a pseudonym $\alpha$ is located often at
a home location and an office location of user *Tom*, then the adversary can
infer with high confidence that the pseudonym a belongs to *Tom*. To prevent
such inference of real identities from pseudonyms, pseudonyms need to be
changed from time to time. However, simply changing the pseudonyms in a
user's path of travel can leave the traces of the user trajectory, and therefore
the linking between the old and new pseudonyms can be easily inferred
using a simple connect-the-dots approach.

Mix-zones securely enable users to change pseudonyms in an anonymous
fashion such that the linking between the new and old pseudonyms cannot
be inferred. The anonymity in mix-zones is guaranteed by enforcing that a
set of users enter, change pseudonyms, and exit a mix-zone in a way such
that the mapping between their old and new pseudonyms is not revealed
(Beresford & Stajano, 2003). A mix-zone of $k$ participants refers to a $k$-ano-
nymization region in which users can change their pseudonyms such that
the mapping between their old and new pseudonyms is not revealed. In a
mix-zone, a set of $k$ users enter in some order and change pseudonyms, but
none leave before all users enter the mix-zone. Inside the mix-zone, the users

do not report their locations, and they exit the mix-zone in an order different from their order of arrival, thus providing unlinkability between their entering and exiting events.

The properties of a mix-zone can be formally stated as follows:

- *Definition 1:* A mix-zone Z is said to provide *k*-anonymity to a set of users A if the following:

  1. The set A has k or more members, that is, |A| ≥ k.

  2. All users in A must enter the mix-zone Z before any user i ∈ A exits. Thus, there exists a point in time where all k users of A are inside the zone.

  3. Each user i ∈ A, entering the mix-zone Z through an entry point e ∈ E and leaving at an exit point o ∈ O, spends a completely random duration of time inside.

  4. The probability of transition between any point of entry to any point of exit follows a uniform distribution; that is, a user entering through an entry point, e ∈ E, is equally likely to exit in any of the exit points, o ∈ O.

Figure 11.1 shows a mix-zone with three users entering with pseudonyms *a*, *b*, and *c* and exiting with new pseudonyms *p*, *q*, and *r*. Given any user exiting with a new pseudonym, the adversary has equal probability of associating it with each of the old pseudonyms *a*, *b*, and *c*, and thus this example mix-zone provides an anonymity of *k* = 3. The uncertainty of an adversary to associate a new pseudonym of an outgoing user *i′* to its old pseudonym is captured by entropy *H*(*i′*), which is the amount of information required to break the anonymity.

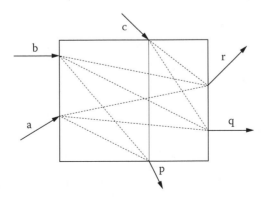

**FIGURE 11.1**
Mix-zone model.

$$H(i') = - \sum p_{i' \to j} \times \log_2 (p_{i' \to j})$$

$$j \in A$$

where $p_{i' \to j}$ denotes the probability of mapping the new pseudonym, $i'$, to an old pseudonym, $j$. When users change pseudonyms inside mix-zones along their trajectories, an adversary observing them loses the ability to track their movements. Mix-zones can be managed by a trusted third party that is independent of LBS providers and mobile users. One such third-party player will be the mobile networking service provider as mobile users use their cell phones to request LBSs through location privacy-protected channels via the networking service provider.

However, several factors impact the effectiveness of mix-zone approaches, such as user population, mix-zone geometry, location sensing rate and spatial resolution, spatial and temporal constraints on user movement patterns, and semantic continuity of the information requested by LBSs. Mix-zones can be constructed inside the building in the walkways and travel path intersections, as shown in Figure 11.2. When mix-zones are constructed on walkways and travel paths inside the buildings, they can be vulnerable to timing and transition attacks because of the inherent nature of the mobility patterns of the users. Concretely, the timing information of users' entry and exit into the mix-zone provides information to launch a timing attack, and the nonuniformity in the transitions taken at the road intersection provides valuable information for a transition attack. These attacks can also be aided by additional background knowledge information of the users available to the attacker. Both of these attacks aid the attacker in guessing the mapping between the old and new pseudonyms. In addition, mix-zones are also prone to CQ attacks when the mobile clients obtain continuous query services. The CQ attack refers to the risk that an adversary can perform inference attacks by correlating the semantic continuity in the time series of query evaluations of the same CQ and the inherent trajectory of locations. Basically, the construction and deployment of mix-zones are challenged by the above-mentioned factors and attacks.

Most existing work on mix-zones has focused on constructing mix-zones on road networks and has not investigated in detail their applications for indoor navigation tools. The idea of building mix-zones at road intersections was first proposed in Freudiger et al. (2007) and Buttyan et al. (2007). An optimal placement of mix-zones on a road map was formulated in Freudiger et al. (2009). These earlier techniques for road network mix-zones follow a straightforward refinement of basic mix-zones (Beresford & Stajano, 2003) by using rectangular- or circular-shaped zones. Both the definition and the construction methodologies of these mix-zones fail to take into account the effect of timing and transition attacks. The road network and CQ-attack-resilient mix-zone framework discussed in Palanisamy and Liu (2011) is the first road-network-aware attack-resilient mix-zone that guarantees an

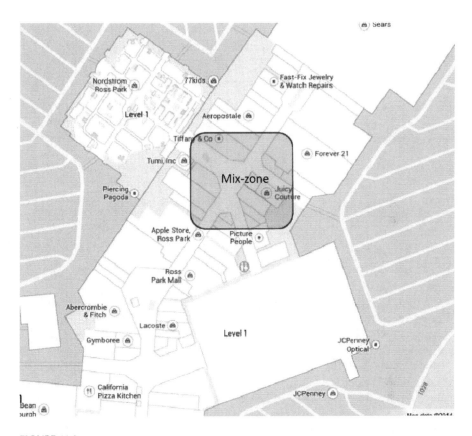

**FIGURE 11.2**
An indoor mix-zone.

expected value of anonymity by leveraging the characteristics of both the underlying road network and the motion behaviors of users traveling on spatially constrained road networks.

The delay-tolerant mix-zones proposed in Palanisamy et al. (2013) combine mix-zone-based identity privacy protection with location mixing to achieve high anonymity for users that is otherwise not possible with conventional mix-zones. In the delay-tolerant mix-zone model, users expose spatially or temporally perturbed locations outside the mix-zone area. However, on the exit of each delay-tolerant mix-zone, the mix-zone changes their perturbed locations by introducing a random temporal shift (temporal delay-tolerant mix-zones) or a random spatial shift (spatial resolution-tolerant mix-zones) to their already perturbed locations. While conventional mix-zones only change pseudonyms inside them, the additional ability of delay-tolerant mix-zones to change and mix user locations brings greater opportunities for creating anonymity. Therefore, the anonymity strength of delay-tolerant

**TABLE 11.1**

Conventional Road Network Mix-Zone

| User | $t_{in}$ | $t_{inside}$ | $t_{out}$ |
|------|------|------|------|
| $o$ | 94 | 4 | 98 |
| $n$ | 96 | 4 | 100 |
| $m$ | 98 | 4 | 102 |
| $a$ | 100 | 4 | 104 |
| $b$ | 101 | 4 | 105 |
| $c$ | 103 | 4 | 107 |
| $d$ | 103 | 4 | 107 |
| $e$ | 106 | 4 | 110 |
| $f$ | 108 | 4 | 112 |

mix-zones comes from a unique combination of both identity mixing and location mixing.

We first illustrate the concept of delay-tolerant mix-zones with an example of a temporal delay-tolerant mix-zone. Table 11.1 shows the entry and exit time of users in a conventional mix-zone. We find that user $a$ enters the mix-zone as $t = 100$ and exits at time $t = 104$. Similarly, the other users enter and exit as shown in Table 11.1. Here the adversary may know that the average time taken by the users to cross the mix-zone is 4 s. Therefore, when user $a$ exits at $a'$ at time $t = 104$, the attacker can eliminate users $e$ and $f$ from consideration as they have not even entered the mix-zone by the time user $a$ exits. Similarly, the adversary can eliminate users $o$ and $n$ from consideration based on timing inference that users $o$ and $n$ have exited the mix-zone by the time $a$ and $b$ enter the mix-zone. Therefore, when $a$ exits as $a'$, the attacker has uncertainty only among the users $\{a, b, c, d, m\}$. Also, among the users $\{a, b, c, d, m\}$, the attacker can eliminate more users through sophisticated reasoning based on timing inference.

However, in the delay-tolerant mix-zone model, each user uses a temporal delay, $dt$, within some maximum tolerance, $dt_{max}$. Inside the mix-zone, the temporally perturbed location of each user is assigned a random temporal shift. In the delay-tolerant mix-zone example shown in Table 11.2, we find that user $a$ initially uses a temporal delay, $dt_{old}$, of 4 s, and inside the mix-zone it is shifted randomly to 16 s. Here $dt_{max}$ is assumed as 20 s. Therefore, when user $a$ exits as $a'$, it becomes possible that many users can potentially exit in the exit time of user $a$. The example in Table 11.2 shows one possible assignment of new temporal delays, $dt_{new}$ for other users in order for them to exit at the same time as $a'$. Thus, during the exit of user $a$ as $a'$, the attacker is confused to associate the exiting user $a'$ with the members of the anonymity set $\{a, b, c, d, e, f, g, h, i, j, k, m, n\}$. In principle, users' new temporal delays, $dt_{new}$, are randomly shifted inside the mix-zone, ensuring the possibility of each of the users to exit at the exit time of each other. Thus, the delay-tolerant mix-zone

**TABLE 11.2**

An Example Temporal Delay-Tolerant Mixing

| User | Observed $t_{in}$ | $t_{inside}$ | $dt_{old}$ | $dt_{new}$ | Observed $t_{out}$ |
|------|------|------|------|------|------|
| $w$ | 81 | 4 | 4 | 20 | 105 |
| $v$ | 84 | 4 | 7 | 20 | 108 |
| $u$ | 84 | 4 | 7 | 20 | 108 |
| $s$ | 87 | 4 | 10 | 20 | 111 |
| $r$ | 89 | 4 | 12 | 20 | 113 |
| $q$ | 90 | 4 | 13 | 20 | 114 |
| $p$ | 92 | 4 | 15 | 20 | 116 |
| $o$ | 94 | 4 | 18 | 20 | 118 |
| $n$ | 96 | 4 | 19 | 20 | 120 |
| $m$ | 98 | 4 | 20 | 18 | 120 |
| $a$ | 100 | 4 | 4 | 16 | 120 |
| $b$ | 101 | 4 | 4 | 15 | 120 |
| $c$ | 103 | 4 | 7 | 13 | 120 |
| $d$ | 103 | 4 | 7 | 13 | 120 |
| $e$ | 106 | 4 | 10 | 10 | 120 |
| $f$ | 108 | 4 | 12 | 8 | 120 |
| $g$ | 109 | 4 | 13 | 7 | 120 |
| $h$ | 111 | 4 | 15 | 5 | 120 |
| $i$ | 113 | 4 | 18 | 3 | 120 |
| $j$ | 115 | 4 | 19 | 1 | 120 |
| $k$ | 117 | 4 | 20 | 0 | 120 |
| $l$ | 118 | 4 | 20 | 0 | 121 |

model provides significantly higher anonymity compared to conventional mix-zones.

A closely related problem to mix-zone construction is mix-zone placement. Although individual mix-zones are efficient with respect to providing the required level of anonymity, careful deployment of them is crucial to ensure good cumulative anonymity for users as they traverse through multiple mix-zones on their trajectories. Mix-zones placed too far from each other may lead to longer distances between adjacent mix-zones in users' trajectories. On the other hand, if mix-zones are placed too close to one another, users may go through mix-zones more frequently than necessary. An optimal solution to the mix-zone placement problem is shown to be NP-complete (Freudiger et al., 2009). Liu et al. (2012) presented an optimal solution to the mix-zone placement problem, which is NP-hard, and presented approximations by relaxing the assumption of nonuniform traffic in road networks. Jadliwala, Bilogrevic, and Hubaux (2011) proposed a game-theoretic approach to mix-zone placement with the assumption that at least one end of each road segment has a mix-zone. Unlike these mix-zone placement techniques that lead to having 50% road junctions as mix-zones, the MobiMix

mix-zone placement techniques proposed in Palanisamy and Liu (2014) are closely integrated with their attack-resilient mix-zone construction methodologies and thereby achieve good privacy even with as few as 10% mix-zones on the road network. Similar to mix-zone construction research, the problem of indoor mix-zone placement has not been investigated yet. We believe that indoor mix-zone deployment is an important and challenging direction of future research and is vital to realizing mix-zone-based location privacy solutions for indoor navigation systems.

## 11.5 Other Location Privacy Protection Techniques

In this section, we briefly review other existing schemes for location privacy protection and discuss their relative merits and demerits. Inspired by the mix-zone concept, there have been several techniques that try to create unlinkability in the user trajectories. The Cachecloak algorithm (Meyerowitz & Choudhury, 2009) employs an alternate technique for path mixing by using cache prefetching to hide the exact location of a mobile user by requesting the location data along an entire predicted path. Although these techniques are effective when all users obtain the same service, they are vulnerable to continuous query correlation attacks when the mobile users obtain uniquely different CQ services.

Recently, content caching (Amini et al., 2011) has been proposed as an alternate solution to location privacy. Here, caching large amounts of information on tiny mobile devices may not be effective. In addition, they may limit the usability of the services by restricting mobile clients to ask only services that are cached beforehand. Shokri, Papadimitratos, Theodorakopoulos, and Hubaux (2011) proposed a collaborating strategy where users can have their LBS queries answered by nearby peers and thereby minimize the exposure of location information to the untrusted LBS. Adding dummy queries to the user's actual queries might help to confuse the adversary about the actual user location. But generating effective dummy queries that divert the adversary is a difficult task, as they need to look like actual queries over space and time. An optimum algorithm for generating dummy queries is an open problem (R. Chow & Golle, 2009). Many techniques have been developed, including adding noise, quantizing locations (essentially putting locations into buckets or aligned onto a grid), and adding false locations. SybilQuery is a client-side tool that creates many different queries to the server to obfuscate the user's actual path (Shankar, Ganapathy, & Iftode, 2009). Hiding a query from the server minimizes the revealed user information and, hence, maximizes his or her privacy with respect to that query. It is more effective than the other three privacy protection methods, and it protects users against both presence and absence disclosures. MobiCrowd (Shokri, Papadimitratos, et al., 2011) hides queries from the server while receiving the query responses from other peers.

Another novel direction of location privacy research is represented by obfuscation-based techniques that change location information by considering the accuracy of location measurements and the privacy requirements (Ardagna, Cremonini, Vimercati, & Samarati, 2011). An alternative thread of research is represented by the *private information retrieval* (PIR) techniques as an alternate to location cloaking for anonymous query processing (Ghinita, Kalnis, Kantarcioglu, & Bertino, 2011; Ghinita, Kalnis, Khoshgozaran, Shahabi, & Tan, 2008). PIR techniques guarantee privacy of mobile users regardless of which types of queries (continuous or snapshot) they ask. However, PIR-based solutions are known to be expensive in both computation and storage overheads, even with the recent new techniques such as hardware-assisted PIR techniques (Williams & Sion, 2008) developed to improve the scalability and efficiency of the PIR approach. Another general issue with PIR-based solutions is their limitation in terms of what kinds of queries can be protected under PIR (Wang & Liu, 2010).

## 11.6 Summary

Advances in sensing and positioning technology, fueled by wide deployment of wireless local area networks, have made many devices location-aware. However, privacy concerns remain the major consumer pushback for a wider acceptance of such services among users. This chapter investigated the location privacy attacks and the protection mechanisms for indoor LBSs. We discussed the various kinds of location privacy and content privacy attacks and reviewed two classes of commonly employed solutions, namely, anonymity-based location cloaking and mix-zone-based location privacy protection. We described the various state-of-the-art spatial and temporal cloaking techniques and mix-zone-based location privacy protection to counter against location privacy threats. While most existing solutions were devised for protecting location privacy of users traveling over road networks, we highlighted the key challenges in adopting existing techniques in the context of indoor LBSs.

## References

Aggarwal, C. (2005). On *k*-anonymity and the curse of dimensionality. In *Proceedings of the 31st international conference on very large data bases* (pp. 901–909). VLDB Endowment.

Amini, S., Lindqvist, J., Hong, J., Lin, J., Toch, E., & Sadeh, N. (2011). Caché: Caching location-enhanced content to improve user privacy. In *Mobisys*.

Ardagna, C., Cremonini, M., Vimercati, S., & Samarati, P. (2011). An obfuscation-based approach for protecting location privacy. In *IEEE Transactions Dependable and Secure Computing*, 8(1), 13–27.

Authorities: GPS systems used to stalk woman. (2002). *USA Today*. Retrieved from http://usatoday30.usatoday.com/tech/news/2002-12-30-gps-stalker_x.htm

Bamba, B., Liu, L., Pesti, P., & Wang, T. (2008). Supporting anonymous location queries in mobile environments with PrivacyGrid. In *Proceedings of the 17th international conference on World Wide Web* (pp. 237–246). New York, NY: ACM.

Beresford, A., & Stajano, F. (2003). Location privacy in pervasive computing. *Pervasive Computing, IEEE*, 2(1), 46–55.

Buttyan, L., Holczer, T., & Vajda, I. (2007). On the effectiveness of changing pseudonyms to provide location privacy in VANETs. In *Proceedings of the 4th European conference on security and privacy in ad-hoc and sensor networks* (pp. 129–141). Berlin, Heidelberg, Germany: Springer-Verlag.

Chen, A. (2010, September). GCreep: Google engineer stalked teens, spied on chats. *Gawker*. Retrieved from http://gawker.com/5637234//

Chow, C.-Y., & Mokbel, M. F. (2007). Enabling private continuous queries for revealed user locations. In *Proceedings of the 10th international conference on advances in spatial and temporal databases* (pp. 258–273). Berlin, Heidelberg, Germany: Springer-Verlag.

Chow, R., & Golle, P. (2009). Faking contextual data for fun, profit, and privacy. In *Proceedings of the 8th ACM workshop on privacy in the electronic society* (pp. 105–108). New York, NY: ACM.

Cuellar, J. R., Morris, J. B., Mulligan, D. K., Peterson, J., & Polk, J. (2003). Geopriv requirements. In *IETF Internet Draft*.

Freudiger, J., Raya, M., Flegyhazi, M., Papadimitratos, P., & Hubaux, J.-P. (2007). Mix-zones for location privacy in vehicular networks. In *WiN-ITS*.

Freudiger, J., Shokri, R., & Hubaux, J.-P. (2009). On the optimal placement of mix zones. In *Proceedings of the 9th international symposium on privacy enhancing technologies* (pp. 216–234). Berlin, Heidelberg, Germany: Springer-Verlag.

Gedik, B., & Liu, L. (2005). Location privacy in mobile systems: A personalized anonymization model. In *Proceedings of 25th IEEE international conference on distributed computing systems* (pp. 620–629). New York, NY: IEEE.

Ghinita, G., Kalnis, P., Kantarcioglu, M., & Bertino, E. (2011). Approximate and exact hybrid algorithms for private nearest-neighbor queries with database protection. *GeoInformatica*, 15(4), 699–726.

Ghinita, G., Kalnis, P., Khoshgozaran, A., Shahabi, C., & Tan, K. (2008). Private queries in location based services: Anonymizers are not necessary. In *Proceedings of the 2008 ACM SIGMOD international conference on management of data* (pp. 121–132). New York, NY: ACM.

Ghinita, G., Kalnis, P., & Skiadopoulos, S. (2007). PRIVE: Anonymous location-based queries in distributed mobile systems. In *Proceedings of the 16th international conference on World Wide Web* (pp. 371–380). New York, NY: ACM.

Gruteser, M., & Grunwald, D. (2003). Anonymous usage of location-based services through spatial and temporal cloaking. In *Proceedings of the 1st international conference on mobile systems, applications and services* (pp. 31–42). New York, NY: ACM.

Jadliwala, M., Bilogrevic, I., & Hubaux, J.-P. (2011). Optimizing mixing in pervasive networks: A graph-theoretic perspective. In *Proceedings of the 16th European conference on research in computer security* (pp. 548–567). Berlin, Heidelberg, Germany: Springer-Verlag.

Karger, P., & Frankel, Y. (1995). Security and privacy threats to its. In *Proceedings of the second world congress on intelligent transport systems*.

Krumm, J. (2007). Inference attacks on location tracks. In *Proceedings of the 5th international conference on pervasive computing* (pp. 127–143). Berlin, Heidelberg, Germany: Springer-Verlag.

Liu, X., et al. (2012). Traffic-aware multiple mix zone placement for protecting location privacy. In *INFOCOM, 2012 proceedings IEEE* (pp. 972–980). New York, NY: IEEE.

Machanavajjhala, A., Gehrke, J., Kifer, D., & Venkitasubramaniam, M. (2006). *l*-diversity: privacy beyond *k*-anonymity. In *Proceedings of the 22nd international conference on data engineering* (p. 24). New York, NY: IEEE.

Meyerowitz, J., & Choudhury, R. (2009). Hiding stars with fireworks: Location privacy through camouflage. In *Proceedings of the 15th annual international conference on mobile computing and networking* (pp. 345–356). New York, NY: ACM.

Mokbel, M., Chow, C., & Aref, W. (2006). The new Casper: Query processing for location services without compromising privacy. In *Proceedings of the 32nd international conference on very large data bases* (pp. 763–774). VLDB Endowment.

Palanisamy, B., & Liu, L. (2011). MobiMix: Protecting location privacy with mix-zones over road networks. In *Proceedings of the 2011 IEEE 27th international conference on data engineering* (pp. 494–505). Washington, DC: IEEE Computer Society.

Palanisamy, B., & Liu, L. (2014). Attack-resilient mix-zones over road networks: Architecture and algorithms. *IEEE Transactions on Mobile Computing, PP*(99), 1.

Palanisamy, B., Liu, L., Lee, K., Meng, S., Tang, Y., & Zhou, Y. (2013). Delay-tolerant mix-zones on road network. *Distributed and Parallel Databases*.

Pan, X., Meng, X., & Xu, J. (2009). Distortion-based anonymity for continuous queries in location based mobile services. In *Proceedings of the 17th ACM SIGSPATIAL international conference on advances in geographic information systems* (pp. 256–265). New York, NY: ACM.

Shankar, P., Ganapathy, V., & Iftode, L. (2009). Privately querying location-based services with SybilQuery. In *Proceedings of the 11th international conference on ubiquitous computing* (pp. 31–40). New York, NY: ACM.

Shokri, R., Papadimitratos, P., Theodorakopoulos, G., & Hubaux, J.-P. (2011). Collaborative location privacy. In *IEEE 8th international conference on mobile adhoc and sensor systems* (pp. 500–509). New York, NY: IEEE.

Shokri, R., Theodorakopoulos, G., Le Boudec, J.-Y., & Hubaux, J.-P. (2011). Quantifying location privacy. In *IEEE symposium on security and privacy* (pp. 247–262). New York, NY: IEEE.

Wang, T., & Liu, L. (2009). Privacy-aware mobile services over road networks. *Proceedings of the VLDB Endowment, 2*(1), 1042–1053.

Wang, T., & Liu, L. (2010). Execution assurance for massive computing tasks [Special session on Info-Plosion]. *IEICE Transactions on Information and Systems, E93-D*(6).

Williams, P., & Sion, R. (2008). Usable PIR. In *Proceedings of the network and distributed system security symposium*.

# *Index*